WIRELESS BROADBAND

Praise for *Wireless Broadband: Conflict and Convergence*

In this book, Vern Fotheringham and Chetan Sharma have done a remarkable job in laying out every aspect of this critically important and demanding topic. Read it from cover to cover, enjoy it all, and be satisfied in knowing what every modern planner, manager, and educated citizen should know about the world's future.

Mark Anderson CEO, Strategic News Service and SNS Project Inkwell

This is as comprehensive a book on Wireless Broadband as I have seen. The authors of this book address the most important issues facing the players in the wireless ecosystem in great detail and provide a very thoughtful analysis.

Steve Elfman, President, Sprint Nextel

Vern and Chetan have written an insightful book that takes a refreshing multidimensional approach to the wireless broadband industry.

Dr. Hyun Oh Yoo, CEO SKT Holdings America Inc. (former CEO of SK Communications)

A comprehensive and insightful coverage of the complex factors affecting the growth of the wireless broadband industry.

Dr. Nitin J. Shah, Cofounder, Personal Broadband Industry Association

This book accurately captures both the historical and holistic perspective on key market forces at work and their interactions that will lead to fulfilling the broadband connectivity requirements of the market.

Umesh Amin, President, Wireless, Intellectual Ventures

With analytic rigor, in-depth analysis, and practical insight, the authors of this book explore the critical aspects of the wireless broadband industry. A must-read.

Dr. Monica Paolini, President, Senza Fili Consulting

The most lucid view yet taken of the confusing world of cellular technology, written by two masters in their field of expertise. If you are in need of education in this area and are searching for the best exposition, you need look no further than this book.

Lowell Tuttman, Universal Consulting Partners

This book is a great overview of the issues relating to our industry and should be of significant value to readers who are curious about how we got where we are, and the lessons we "should have" learned along the way as an industry whose fate was and is tied so closely to the risk capital market.

Bill Rouhana, former CEO and Chairman, Winstar

WIRELESS BROADBAND

Conflict and Convergence

Vern Fotheringham
Chetan Sharma

IEEE PRESS

A JOHN WILEY & SONS, INC., PUBLICATION

Library of Congress Cataloging-in-Publication Data is available.

ISBN 978-0470-22762-6

Printed in the United States of America

10 9 8 7 6 5 4 3 2 1

CONTENTS

7 THE EMERGING INFLUENCE OF THE COMPUTER INDUSTRY 135

8 ALWAYS BEST CONNECTED 145

FOREWORD

There is probably something in the world of technology more important than the subject of wireless broadband, but I'm pressed to think what it is.

Wireless broadband is the point of a spear which, in every country in the world, will drive progress in education, economic development, health and medicine, agriculture, markets, family welfare, technological and scientific advances, and general communications.

One can say that these things are equally true of wired broadband, but it's more true on paper than on the ground. Or, in more stark terms, wired broadband, while it offers fatter pipes today via glass fiber, is generally for the urban well-to-do. And even in enlightened countries which have a more socialized view of bandwidth (Iceland leaps to mind, as well as Japan, South Korea and Sweden), the problem of rural delivery remains.

In fact, we seem, as a planet, to be on the verge of a mammoth deployment of bandwidth, and my guess is that the great preponderance of those cycles will be delivered wirelessly. Today, we see this in Western cities as a trend in voice from wireline to wireless which is breathtaking at the carrier boardroom—and Wall Street analyst—level. At the same time, this impending revolution is clear from the low cost, rapidity and appetite for wireless broadband deployments in the poorest and most rural environments, from Vietnam to India to Africa.

In general, I've been willing to make a rather treacherous bet: that a correlation will emerge, after a decade or two, between a country's deployment of bandwidth (technically, its penetration), and its economic growth. As I write this, Kevin Rudd has just won the office of Prime Minister in Australia on a platform that includes widescale provision of broadband, and what we at SNS Project Inkwell call one to one computing in schools, where each child has a Net-connected computer. At the same time, the United States is in the midst of a presidential race, and neither candidate has mentioned bandwidth, and I strongly doubt they will.

Yes, my economic bet is on Australia in that race. So this is not purely a technical discussion or issue, and in my opinion, those who take the subject to heart, and get out ahead early, will be the winners, almost regardless of individual pursuit or other national agendas.

There is a rather fascinating aspect of bandwidth consumption which further tilts the bandwidth table in favor of wireless: carriers and everyone else in the technology world consistently underestimate even the near-term future demand for bandwidth. "I'm running as fast as I can," seems to be the basic point of view expressed by most carriers, but most of them are about to get a very rude awakening, brought on by accelerated demand by their customers and accelerated provision by their competitors.

A few years ago, providing residential customers with 256 thousand bits per second through any medium was considered adequate; today, in downtown Paris, 12 million is normal. A few years ago, families in the United States were used to spending $50 per month for telecommunications needs. Today, those same families are spending twice that and more for bundles that include Internet and voice communications, video entertainment, gaming, and other services, often reallocating budgets in ways that have caught most providers by surprise.

How much bandwidth will be enough tomorrow? Five years ago, the Canadian province of Saskatchewan—a global leader in bandwidth—did internal estimates that 25 Mbps would be enough per home. Today that figure seems shy. Yesterday, a single video stream used about 1.5 Mbps; today, high-definition TV runs around 14 million, and Asian companies are making TVs that sample at half again higher rates. How many independent screens will there be per home? In many cities, the idea of independent viewing in the bedroom, living room, den, and kitchen is not far-fetched, but suddenly we're talking about a 60 Mbps home.

Don't use such skinny numbers around my friend Larry Smarr, however. Larry is the founder and director of the world's most advanced visualization laboratory, CalIT2, and he regularly tells our Future in Review conference participants that 100 Gbps (billion bits per second) is about the right number. Of course, Larry is driving a wall-sized screen with 220 million pixels (dots), but he'd tell you that you, too, will have one someday, or something as close to it as you can afford.

One more example of near-term unfulfilled demand will help illustrate the dramatic importance of wireless broadband. Consider K12 education, which promises to become the largest market segment for computers sometime during this next decade. While everyone is wondering where the funds will come from for one computer for each student (and teacher), most planners are overlooking a more important question: bandwidth.

How much does one student need? Do they want to watch movies? Of course! Well, that's about 1.5 Mbps. Does the teacher want them to be able to watch the same movies as the other children in the class? Of course! How many kids in the room, maybe 30? All right, that's 45 Mbps. And how many classrooms in the building? Perhaps 15 or more, plus a library, assembly room, etc.; perhaps the building needs 675 Mbps. Whoa! How do you get it, and who is going to pay for it? This may be the largest problem facing modern elementary education today.

It is this insatiability for cycles which puts wireless in the foreground: wires (and fiber) just can't keep up. For the moment, and as long as fiber remains the fatter pipe, one can picture the globe as though two kinds of wildfire were consuming it: first comes the wireless provision, followed in the cities by the wired provision. If wireless becomes the fatter pipe—and there are reasons to think this could happen—all fiber bets are off.

As though the global trends named here were not sufficient drivers to warrant attention to wireless broadband, there is another, equally compelling set of accelerants, all coming under the umbrella title of mobility. On every level, from lifelong residence to lifestyle to work, humans are becoming more mobile by the decade, and wireline bandwidth, while growing, is increasingly not appropriate to our needs. Cars today have more computers in them than houses, but get a small fraction of the comparative bandwidth. That will change.

Finally, it is worth noting that wireless bandwidth will be the Great Equalizer of this century, providing citizens and countries equal access to the world's information and commerce. Countries which, like China (yet to move to 3G), choose to put politics ahead of this trend, will become case examples of what not to do, while those such as India which push aggressively for wireless bandwidth will be emulated worldwide, for the hope and prosperity which this form of being connected can bring.

In this book, Vern Fotheringham and Chetan Sharma have done a remarkable job in laying out every aspect of this critically important and demanding topic. Read it from cover to cover; enjoy it all; and be satisfied in knowing what every modern planner, manager, and educated citizen should know about the world's future.

MARK ANDERSON
CEO, Strategic News Service
and SNS Project Inkwell

Friday Harbor, Washington State
August 2008

ACKNOWLEDGMENTS

Periodically life provides the privilege of meeting one of those special people who make regular substantive contributions to enhancing their chosen field of expertise, but who do so with a modesty and humility that is simultaneously both engaging and completely natural. Chetan Sharma is one of those special people possessing a profound competence and a true global view. Chetan and I have been acquainted for several years, but we never had an opportunity to collaborate directly until Chetan suggested we co-author this book. I had often contemplated such a project, but had no prior experience with the publishing industry and the many aspects that were involved in bringing a book from idea to publication. I thank Chetan for the opportunity to work with him, and many benefits enjoyed from his mentorship. This project has brought me to a heightened respect for the efforts and commitment required by authors to translate their ideas and experience into the printed word. I will never be able to walk through a bookstore in the future without an overwhelming sense of respect and admiration for the sheer amount of effort represented by all those volumes.

I am also deeply grateful to the huge number of former colleagues, customers, investors, and vendors who have contributed to my awareness and knowledge about so many facets of the industry. It has been a true privilege to have been exposed to so many dedicated and knowledgeable people who have shaped my outlook and insights. The nature of this project is such that I must thank virtually everyone who has played a role in providing me the privilege of participating in so many opportunities along the leading edge of the competitive telecommunications industry over the past 25 years. This is a list that is far too extensive for this brief salutation, however, a short list of my many mentors, teachers, supporters, clients, investors, or partners has to include: The Honorable Dennis Patrick, Allen Salmasi, Dr. Irwin Jacobs, Quincy Jones, John O'Steen, Mike Kedar, Ulysses Auger, Trond Johannessen, Arve Johansen, Albert Hawk, Masao Ono, Craig McCaw, John Stanton, Gordon Rock, Richard Munroe, David Arthur, Philip Garrett, Joseph Walter, Ted Pierson, Dennis Burnett, Laurence Zimmerman, Keith Markley, Marc Arnold, Alan Cornwall, Mark Marinkovich, Ted Ammon, Kenji Ishikawa, Governor Mark Warner, Dan Hesse, Keith Grinstein, Christian Seifert, Linda Nordstrom, Dr. Doug Reudink, Dr. Richard Baugh, Robert

Foster, Tom Huseby, Dale Miller, Jim Miller, Jamie Howard, Ambassador Bradley Holmes, Charlie Schott, Charles Menatti, Jerry Cady, Michael Cote, Dr. Farshad Mohamedi, Bob Nitschke, Jim Parsons, Jim Frank, Wolfgang Mack, Dr. Hui Liu, Dr. Xiaodong (Alex) Li, Vern Stevenson, Rob Manning, Brad Knight, William Cortes, Esq., Bob Beran, Dr. Fumio Murakami, John Humphrey, Brooks Harlow, Dr. C.R. Baugh, Lowell Tuttman, and the many others who have contributed to my unique and expansive experience.

Specific thanks for help in bringing this book into reality must extend to my circle of peer reviewers and informal editors who plowed through some pretty ponderous drafts to help me focus the work and avoid factual and stylistic mistakes. Special thanks to Ted Pierson and John Humphrey for their contributions and to the IEEE assigned peer reviewers each of whom took careful diligence to review and comment on the work through various drafts.

It is also imperative that I thank my children, Brooke, Graham, John and David, who have all sacrificed and supported my adventures in business and life. They have grown into remarkable and creative individuals who provide me a constant source of pride and gratitude for their daily demonstration of character and humanity. Their lives reflect so well their mother's nurturing and guidance, successfully offsetting the fact that their father was so often absent during their formative years while pursuing his obsessions with business and learning. I thank you all with a deep sense of humility and gratitude from the depths of my spirit.

V. F.

I thank my co-author Vern. It has indeed been an honor working with someone as knowledgeable as Vern. His entrepreneurial zeal is quite contagious and I treasure the book review meeting in Beijing as we crisscrossed the globe during the course of this project.

I thank clients, friends, and colleagues of Chetan Sharma Consulting who have helped shape my thinking over the years and had a profound impact on the outcome of this book. Many thanks to Umesh Amin, Steve Elfman, Joe Herzog, Sunil Jain, Dave Smiddy, Marianne Marck, Ike Lee, Mark Anderson, Dr. Hyun Oh Yoo, Dr. Nitin Shah, Monica Paolini, Brendan Benzing, Victor Melfi, Subba Rao, Frank Barbieri, Dev Gandhi, Dr. Young-Chu Cho, Dr. Yasuhisa Nakamura, Om Malik, Tomi Ahonen, Mike Vanderwoude, Dmitry Kaplan, Mat Hans, Scott Weller, Akio Orii, Mitul Patel, Linda Liu, Michel Gaultier, Paul Palmieri, Sami Muneer, Rajeev Agarwal, Oscar Alcantara, Brian Fagel, Mani Prakash, Bruce Grant, Tom Patterson, Brian Vincent, Elizabeth Aleiner, Dave Keller, Mario Obeidat, and countless others who have been generous with their time and intellect over the years.

Our sincere thanks to Mark Anderson for taking the time to pen an excellent Foreword for the book. He has been a source of inspiration to many for many years.

I owe the most to my family. No book project can be successful without the selfless sacrifice of loved ones. I thank my parents Dr. C. L. Sharma, Prem Lata

Sharma, and Dropadi Sharma who have instilled in me the desire for hard work and honesty; brother-in-law Aditya, and brother Rahul for their support and encouragement of whatever I pursue. I thank my better half—Sarla—for making my life *truly* better. She has supported me through all book projects and have never wavered in enthusiasm, encouragement, patience, and understanding. Writing a book is never easy on the family but Sarla's support has always made it so easy for me. And finally, I thank Maya, my 4.5-year-old angel, who accommodates her Dad's long hours with a smile and enduring patience and my newborn Anish who is looking forward to his Dad's long hours.

C. S.

We also thank the readers who picked up the book and hope to continue the conversation. Please feel free to contact us at vfotheringham@yahoo.com and chetan@chetansharma.com.

LIST OF FIGURES

INTRODUCTION

The telecommunications industry has evolved to a point in time when the wireless elements of the global network have eclipsed the legacy wired networks in terms of reach and adoption by the world's population. There is now a growing tension between the original vision of the cellular network as simply a mobile extension of the traditional wired telephone network that is operated as a closed system under the unilateral control of the service provider and its role as a leading access platform for the global Internet. The powerfully established business and regulatory model of the legacy telephone network operators is now bumping up against the dramatic expansion of the global Internet into a broadband data system that can provide alternatives for virtually every legacy communications service. A historic conflict is evolving over how these two mammoth environments will converge and overlap. Will the well-established institutions that hold sway over the legacy telecommunications networks and service providers capture control of the Internet by leveraging their existing gatekeeper position for access and termination? Alternatively, will these well-established habits of operation yield to creative new forces and competitors who will grow and thrive by implementing new business models that make obsolete the business practices of the incumbents? This conflict is well under way, and its outcome will have tremendous influence on the future of the global economy, the evolution of human rights and freedom, and the daily lives of virtually all the world's citizens.

The core theme of this book is an examination of contesting factors that have influenced and will continue to influence the deployment and adoption of the broadband Internet Protocol (IP) wireless infrastructure, its devices and its services, which will mark the next major steps in the evolution of wireless worldwide. The implementation of the ubiquitous wireless broadband Internet will reach into every corner of global society. Every segment of the wireless industry will ultimately have to view and plan for its future prospects from the perspective of how it will fit within the emerging IP ecosystem growing out of this major change of state for the entire telecommunications industry. We will consider the impact of new entrants and operators, versus new innovators and the current market leaders in each sector of the industry. We will also examine how the future technology road maps of the 3GPP (Third Generation Partnership Project) and WiMAX (Worldwide Interoperability for Microwave Access) standards promoters will conflict, compete, and ultimately converge. Our efforts will also seek to penetrate the noise and hype, both positive and negative, that presently cloud the perceptions of both industry insiders and the larger publics who will be impacted by this insidious and inevitable broadband evolution.

New broadband wireless deployments will find market share both among and beyond the current base of 3 billion subscribers, most of whom are on second generation (2G) versions of the global system for mobile communications (GSM) systems. The installed base of GSM infrastructure is presently undergoing a slower than anticipated, but inevitable transition to third-generation (3G) platforms. This step along the trajectory to true broadband IP–centric fourth-generation (4G) networks can be viewed as the transition from the narrowband 2G environment to the wideband 3G era, which will evolve into the true broadband future matching the vision of the 3GPP technology Long Term Evolution (LTE) for GSM systems, and the emergent Mobile WiMAX standards based on OFDMA (orthogonal frequency division multiple access) technology. The emergence of OFDMA as the technology of choice for the next-generation mobile platforms is a by-product of the dramatic increases in microprocessor power over the past decade that finally enabled OFDMA technology to become practical for application in wireless platforms. These systems will come into existence under the sponsorship of existing cellular operators, and through major telecommunications and computing industry organizations that have to date been essentially left out of direct participation in the wireless industry. Included among these new contributors to the wireless broadband future are the cable television operators, Internet portal and search companies, computer and digital appliance manufacturers, software concerns, and content developers.

Much of the momentum driving mobile wireless broadband services is being created by the widespread adoption of wired broadband Internet services by a large portion of the population. The experience and convenience of broadband access have extended from their original presence in the workplace into approximately 60% of all U.S. households, primarily though digital subscriber line (DSL) and cable modem services.* We are now at the tilting point when it is both practical and logical to seek access to our broadband services and applications wherever we may be, regardless of whether we are at the office, at home, traveling to a remote destination, or mobile betwixt and between these locations. We will address the nature of network and service convergence and the interrelationships that exist between and among each of the broadband network service domains, including all types of wired and wireless networks.

There is a pending collision between the traditional telecommunications industry closed system approach to the market and the open platform environment of the Internet. As broadband wireless service delivery networks proliferate, the

---

* World Broadband Statistics—Q2 2007, Point Topic Limited.

migration to expanded openness will accelerate. The traditional "walled garden" environments of the legacy wireless service providers are already breaking down, with pledges to remove existing carrier-defined constraints that only allow network access to user devices obtained from the underlying carrier coming from both Verizon and Mobile WiMAX proponents. How these deeply established traditions of the telecommunications industry are relaxed and eliminated in whole or in part and at what pace over time will mark the next era of the wireless industry.

Numerous contributing factors will impact the pace of the ubiquitous availability of wireless broadband services. These include: (1) the need to resolve a wide range of regulatory constraints and protectionist policies on literally a global basis; (2) the existence of enabling technology development for pending broadband wireless expansion in an increasingly complex intellectual property environment that requires equipment manufacturers to be sensitive to potential business risks, which are very difficult to quantify in advance of drawn-out contentious legal processes; (3) the need for substantial increases in the amount of radio spectrum allocated to existing and new service providers with sufficient contiguous bandwidth to support truly broadband services; and (4) the need for non-discriminatory standardization of networks and user equipment across commercial and political boundaries, which will likely take many years to resolve.

We will attempt to handicap the field contending to be the future winners and losers among the numerous competing factions participating in the broadband convergence movement. Included among the participants for next-generation network services leadership are the reconsolidated and expanded (wireless, Internet, video and long-distance-enabled) legacy telephone companies, called the incumbent local exchange carriers (ILECs), non-ILEC cellular network operators (Cellcos), the multisystem operators (MSOs) in the cable industry, wireless internet service providers (WISPs) led by the new Mobile WiMAX system operators, the direct broadcast satellite (DBS) service providers, and the competitive local exchange carriers (CLEC).

Our direct experience over the past 20 years of the evolutionary march of progress towards a wireless broadband future has revealed many of the obstacles and obstructions that have emerged either as defensive acts of commission by established operators, or acts of omission on behalf of regulators and vendors, which have resulted in a seemingly never-ending series of chicken-or-egg phenomena. Inefficiencies impacting progress abound, including how wireless spectrum is allocated and licensed, how capital formation is organized and aligned with new network requirements, and how the numerous "standard" obstacles that mark the implementation of wireless infrastructure are overcome, such as site acquisition in a crowded market, local zoning obstacles including NIMBY (not in my back yard) issues, and the growing challenge of provisioning broadband backhaul and interconnection for cell sites with vastly increased capacity requirements compared with legacy voice cellular systems.

We are attempting to cover a very wide swath of the issues facing decision makers within the impacted sectors of the economy, with the intention of broadening their awareness of emerging competitive factors and potential opportunities that will decide their future success or failure. In addition, we hope to add worthy contributions to the policy making process to add additional insight and information to the impacted publics on every side of these often polarized issues.

We all share responsibility for the future we create as members of our respective professions and societies as well as members of the global community of nations. Our world is shrinking rapidly, and few technologies are contributing to this evolution of global interaction and interdependency as completely and cogently as broadband communications in all of its multivariate forms.

WHERE WE ARE

Wireless Meets the Broadband Internet

WHERE WE ARE

Today many persons equate the word "wireless" with cellular, forgetting the huge swath of other applications this ethereal technology provides, but responding to the overwhelming success of cellular, which has grown from trial systems in the early 1980s to about 3 billion subscribers globally by the end of 2007. Market researchers are now predicting that the migration from narrowband 2G cellular to 3G wideband voice and data systems and subsequently to 4G networks capable of delivering true broadband mobile services to wireless subscribers will easily exceed 1 billion by 2012 [1].

As widespread interest in personal computing swelled in the early 1980s, it became apparent that complex digital wireless technologies were able to become practical and cost effective as the effects of Moore's law,* which had driven the success of distributed computing, spilled over into the wireless industry. The wireless industry has now emerged as an integral element in the broader digital universe of computing, software, application-specific integrated circuits (ASIC), digital signal processors (DSP), memory and logic processor chips, materials science, automated design and manufacturing tools, and storage. The immediate result of these developments is the increased velocity of product development,

---

*Posited in 1965 by Gordon Moore, the cofounder of Intel. Since the density of transistors on integrated circuits had doubled every year up to that time, Moore's Law stated that this progress would continue into the future, which has largely proven correct. In recent years, the pace of Moore's law has continued, but the doubling is now taking place about every 18 months.

Wireless Broadband. By Vern Fotheringham and Chetan Sharma
Copyright © 2008 the Institute of Electrical and Electronics Engineering, Inc.

the commoditization of components and assembled hardware, and the efficiency and mass specialization capabilities of robotic assembly techniques.

The wireless industry has long been influenced by regulation, often heavy-handed but sometimes inspired, in virtually every nation worldwide. In the United States, at the time of this writing, there are new voices joining the public policy debate. These include the growing influence of Internet software–centric organizations such as Google, which is becoming visible on the public policy stage for the first time. The software content and applications industry has begun to pressure the Federal Communications Commission (FCC), both directly and through initiatives, to gather support in Congress to back the migration to open network access policies for future spectrum licensees.

So where are we on this evolutionary path to the future? In our opinion, we are "muddling through" [2] a period of transition to a more efficient, cost-effective, and flexible wireless broadband future. The forces of inertia, vested-interest protectionism, and alternative technology overload are arrayed against the financial tension between upside opportunities and downside risks at the extreme ends of the scale.

We will all ultimately arrive in a ubiquitous broadband world where we can access every type of service from virtually any location—the proverbial wireless broadband Internet future. How we manage the conflicts between and among the impacted parties will determine how soon the crystal clear vision of a broadband future arrives. Balancing the forces of change against the power of incumbency will require a thoughtful public policy dialogue that results in an efficient and least destructive path forward. We are going to get the future we deserve, but we will certainly deserve what we get, in a negative context, if we fail to approach the invention of our shared future and our position in the global technology market, without due care and awareness of all the issues, both positive and negative.

The present market development environment for next-generation network implementation has been formed by the tortured and financially traumatic events of the past eight years. The competitive telecommunications industry has come through a forge of financial restructurings of unprecedented scale, which has created a strong foundation of low cost facilities and has been complemented by the insidious march of technology development that both enhances network performance and reduces costs.

The following section describes the events that led to the collapse of the industry and the reconsolidation of the incumbent Bell operating companies into competitors that are more powerful than they were as independent ILECs.

HOW WE GOT HERE: REINTEGRATION OF THE TELECOM OLIGOPOLY AND CRACKS IN THE WALLED GARDENS

The Disassembling and Reassembling of the Telecom Industry and the Collapse of Monopolies and of Competition

The global telecommunications market is heading into a dramatic period of change that will result in a significant restructuring and realignment of the economics and financial underpinnings of the industry. The devolution of the monopoly telephone organizations into competitive environments has been progressing steadily over the past 25 years.

In 1982 the Modified Final Judgment (MFJ) [3], a modification to the original 1956 consent decree, between the government and AT&T, after negotiations, was issued by federal judge Harold Greene to settle a suit between AT&T and the Federal Trade Commission. This landmark decision resulted in the breakup of the Bell monopoly into seven regional Bell operating companies (RBOCs). The original 1956 Final Judgment had mandated that

- AT&T be prohibited from entering the computer and information services business
- the Western Electric equipment manufacturer be mandated as a separate subsidiary
- AT&T Long Distance and the local Bell companies be established as subsidiaries
- Bell Labs be separated as Telecommunication Research

The 1982 MFJ called for

- the divestiture of the local Bell operating telephone companies into seven RBOCs
- AT&T's retention of its long-distance (LD) and manufacturing businesses
- The disallowance of RBOCs to manufacture or to get into the LD business
- the prohibition of AT&T against providing local telephone service or acquiring the stock or assets of any of the RBOCs.

The divestiture agreement incorporated a "triennial review" system under which Judge Greene reviewed the evolving competitive telecommunications market and ruled on whether proposed AT&T initiatives or entry into new or legacy market segments would be allowed. These triennial reviews continued well into the 1990s and were essentially rendered obsolete by the Telecommunications Act of 1996.

The original negotiated settlement was ultimately a trade-off between the allocation of assets between the new RBOCs and AT&T. Ultimately, AT&T decided to maintain its powerful position in the LD business, while swapping out the local access, cellular, and yellow pages businesses to the seven RBOCs. It is a widely held belief that the objective of divestiture was primarily to foster competition between and among the RBOCs, which were formidable and sizable local monopoly organizations. The evolution of the market environment, in large part driven by the advent and tremendous growth of the cellular and cable industries, eventually reduced the strict separation under the MFJ that had been overseen by Judge Greene and its relevance faded as market conditions evolved.

However, many of the arbitrary legacy regulations and business line separations wound up being codified in the 1996 Telecom Act.

Emerging competitors persuaded the FCC into believing that the entrenched position of the legacy telephone companies gave them an unfair advantage in the marketplace and that without relief on pricing and access to facilities there would be no realistic chance for a healthy, competitive marketplace. The competitors successfully argued that they could not afford to replicate the telephone industry's end-to-end infrastructure. Consequently, the FCC promulgated a series of decisions that afforded the competitors access to parts of the existing wireline infrastructure at rates that were less than the standard retail rates charged by the ILECs. This culminated in congressional legislation, the 1996 Telecom Act, which enshrined and enlarged on these policies while creating a path for the RBOCs and their sisters in GTE and United Telecom to rejoin, provided they could demonstrate that their markets were on the road to being competitive.

Unfortunately, less than half a dozen years after the 1996 Act, much of the newly created CLEC telephone sector largely collapsed in 2000–2001. The precipitating cause of this collapse was the massive withdrawal of capital markets' interest in funding operating losses. There were a number of reasons for the almost overnight shutdown of access by the CLECs to construction and operating funds. The somewhat irrational contagion from the dot-com meltdown was one reason. It was also true that there had been widespread irrational overbuilding of duplicate fiber-optic networks on the same routes and in the same high density metropolitan areas. Too many CLECs were funded to implement virtually identical business plans in the same markets. A successful business plan, for instance, for long-haul fiber or metropolitan fiber rings was soon copied, and all too often over the same routes or in the same metropolitan area.

There were two significant contributors to the early failures among the CLECs that were common to all of them, regardless of whether they were simply resellers: facilities-based fiber network operators and wireless network operators. We will also address the impact of the behavior of the ILECs after we describe the CLECs that failed due to these factors.

Time Constraints. One problem was the comparatively short period of time available to the CLECs that had entered the market on the premises and promises of the 1996 Telecom Act who were in business before access to the capital markets was shut down. These emergent competitive organizations had only three to five years in which to build their capital-intensive businesses and networks before the market's financial collapse. In most cases, this time was insufficient for the organizations in most cases to complete network builds, to build their customer base to critical mass, and to reach positive cash flow, particularly with the debt that they had to raise in order to build even the comparatively limited amount of infrastructure their business cases required. In sharp contrast, their competitors, the legacy telephone companies, had almost 100 years to leisurely build their networks without any threat from competitive pricing. Indeed ILECs were treated as public utilities entitled to a guaranteed profit on whatever capital they invested in the business. In short, the more money they spent, the more they earned. Thus, they were completely protected from any bad business decisions or poorly conceived investments. The ILECs could invariably tap the equity and debt markets at almost any moment for whatever funds they needed at low interest rates. The CLECs too raised money in the equity and debt markets much as the ILECs did. However, they were usually forced to pay much higher interest rates. For instance, MCI Telecom, which was the first CLEC, was one of the first companies to raise debt through high yield debt securities, the so-called junk bonds pioneered by Michael Milken at Drexel Burnham. The rates on these bonds were two to three times the rates paid by the legacy telephone companies. When the downturn came, the CLECs were the hardest hit. When their stock prices plummeted, they fell into default on the covenants for the high yield debt securities burdening their balance sheets. In most instances, they also had too small a customer base to pay the interest rates and were unable to raise additional equity.

The Virtuous Circle Turned Vicious. The venture capital (VC) and private equity industries, fully aided and abetted by the investment banking industry, were responsible for the early funding of the CLEC industry and hence for their existence. However, their participation all too often came at a price, which contained the seeds of failure. In their frenzy to exact fees from the process and attract investors, the investment banks usually insisted on unrealistic business plans.

In the late 1980s, the common time frame in which a VC sought a return on its investment was five to seven years. By the mid 1990s, this period was drastically truncated to 18–36 months. This meant that a new CLEC was expected to "go public" or be acquired in less than three years from its initial funding. In order to be attractive to the public markets and to third-party acquirers, the CLECs were urged to build rapidly and expansively. For instance, in the case of the three fixed wireless companies that went public in this period (Teligent, WinStar, and Advanced Radio Telecom) and each of which ended up bankrupt, all were pressured to rapidly rollout nationwide deployments. The mantra of the investment banks was, "We only back category leaders and to be a category leader in this industry you have to be national." For an early stage company, no matter how experienced the founders and line management, this was a monumental undertaking not only in terms of capital but in terms of human resources and management systems. At least in retrospect, this approach was doomed to failure, and fail it did devastating the equity values of numerous companies that had invested tens of billions in state-of-the-art new network infrastructure.

Virtually all of the sector leaders went public early with ambitious plans to expand rapidly nationwide, tapping the high yield debt markets to fund their growth, and subsequently had to seek bankruptcy protection and restructuring. This trend resulted in an extraordinary reduction in their equity values, as they were reorganized with the debt holders emerging as the new equity base. Thus, with the bondholders becoming the new equity owners, the resulting restructurings led to a dramatic slowdown in CLEC expansion and operations, which subsequently trickled down to the broader telecommunications support industry, damaging numerous equipment vendors and many professional services organizations that had emerged to support the rapid growth.

A Description of Competitors Who Failed. The leading facilities-based CLECs and (data only) DLECs that were driven into receivership included the following:

1. XO Communications

   The largest surviving CLEC, XO Communications went bankrupt and was reorganized under XO Holdings (listed under OTCBB: XOHO); Carl Ichan now controls it. XO also owns Nextlink Wireless, its subsidiary for wireless broadband access services, which holds the largest footprint of 28 GHz LMDS (Local Multipoint Distribution Services) licenses throughout the United States.

2. McLeod USA

   Headquartered in Cedar Rapids, Iowa, McLeod USA is a CLEC and interexchange carrier (IXC) that provides integrated voice and data services to small and medium sized businesses in a service territory that includes the Midwest, the Rocky Mountain states, and portions of the Southwest and Northwest United States. The company provides traditional local and LD services, and high speed Internet and value-added data services.

3. Network Plus

   Network Plus was acquired out of its prepackaged bankruptcy by Broadview Networks, which continues to use Network Plus to provide integrated service provider operations in the northeastern region of the United States.

4. Mpower

   Mpower was acquired by TelePacific, a California CLEC, which subsequently sold Mpower of Illinois to McLeod USA in May 2007. The combined entity is now being acquired by Paetek, headquartered in Fairport, New York, to add to its existing footprint of CLEC-integrated voice, data, and Internet services.

5. e.spire

   e.spire was acquired out of bankruptcy by Xspedius Communications, LLC, headquartered in O'Fallon, Missouri. Xspedius provides integrated telecommunications services to small to medium sized enterprises (SME) throughout the southern region of the United States. Xspedius was created through the consolidation of several telecommunications companies, including its original operations in Lake Charles, Baton Rouge, Lafayette, Memphis, Nashville, and Greensboro/Winston-Salem. In August 2002, Xspedius purchased substantially all of the assets of Virginia-based e.spire Communications, which added 55 markets across 20 states, and more than 3,500 total route miles of fiber to Xspedius' assets.

In January 2003, the company acquired Mpower Communications' Texas assets in the Dallas/Fort Worth, Houston, Austin, and San Antonio markets. In April 2005, Xspedius purchased the business and assets of ICG Communications Inc. in five markets across the Southeast, enhancing its service in Atlanta, Birmingham, Louisville, Nashville, and Charlotte, NC.

6. Focal Communications
   Focal was on the brink of insolvency when it was sold to Broadwing Communications, the CLEC subsidiary of Corvis Corporation, in September 2004. Subsequently, Level3 purchased Broadwing Communications in January 2007 to expand its portfolio.

7. Broadband Office
   This Kleiner Perkins and real estate industry–backed venture was lavishly funded with approximately $250 million in VC, which it consumed in only about 18 months, while seeking to capture the major office buildings as the "last monopolies" for telecommunications services. It folded in the midst of the CLEC implosion in 2000–2001.

A number of very large Internet service providers (ISPs) were also driven to seek bankruptcy protection, followed by restructuring or liquidation, including

1. PSINet
2. Ardent Communications (CAIS Internet)
3. Excite@Home
4. iBeam
5. NetRail
6. colo.com
7. Exodus

In addition, most of the leading fiber-optic network operators and competitive IXCs were swept into the massive destruction of equity among the facilities-based carriers. The bankruptcies in this sector included

1. 360 Networks
2. Ebone/GTS
3. Global Crossing
4. Storm Telecommunications
5. Enron Broadband
6. MCI-Worldcom

Also pushed into restructuring were all of the leading broadband wireless and several of the leading mobile data and satellite pioneers, including

1. Winstar
2. Teligent

3. Advanced Radio Telecom
4. Formus
5. OmniSky
6. Metricom/Ricochet
7. GlobalStar
8. Iridium

A similar fate befell the competitive DSL access service providers. The competitive DSL providers sought to leverage the "unbundled" copper network elements of the ILECs, which had been made available to them by the 1996 Telecom Act.

Each of the pioneering firms in the DSL space were either forced out of business or forced to reorganize in following widespread telecom sector financial market collapse. The most noteworthy of these were

1. Covad Communications
2. NorthPoint
3. Rhythms NetConnections

In our opinion, the provisions of the 1996 Telecom Act that made the last-mile copper loops of the ILECs available for lease to the DSL providers and that permitted them to install their own digital subscriber line access multiplexers (DSLAMs) in the ILEC central offices well illustrate the law of "unintended consequences." The effect of the Act was to tie the DSL providers inexorably to the ILECs. The competitive DSL companies became utterly dependent on the ILECs' last-mile facilities. This dependence meant that they were subject to whatever pricing and conditions of service that the ILECs were permitted to impose. This in turn meant that when, as was inevitable, the ILECs gained the ability to increase prices, to dictate onerous service conditions, and even to deny them access to certain facilities, the DSL providers' business cases were torpedoed. And the ultimate irony was that the DSL providers initially were flooded with capital that otherwise could have been spent on constructing alternative last-mile facilities, which would have had twin advantages: the consumer, especially the SMEs, would have had a second set of competitive facilities to leverage and the DSL providers would have not been at the mercy of the ILECs.

In addition to these DSL access providers, there were hundreds of ISPs, who were resellers of competitive DSL services, also suffered substantial damage. One of the principal culprits, though, was not just the capital markets or the downturn in customer demand. Rather, many of these ISP resellers suffered as a result of pernicious practices of some of the ILECs. These practices at best amounted to "strategic incompetence."

It would have been manifestly illegal for ILECs of any stripe to refuse to provide access to their copper facilities to the DSL resellers. It was, however, not illegal for the ILECs to provide poor service to the resellers who were competing with them, but who were nevertheless dependent on them, or at least, it was much more difficult for the competing resellers to prove the existence of a sufficiently pervasive and systematic pattern of poor service to amount to an illegal practice. The opportunity for harmful activities arose from the need for the ILEC as the wholesale provider to continually take actions in cooperating and coordinating with the reseller to cut over service to customers switching from the ILEC to the reseller.

The ILEC actions included such "techniques" as being late for appointments, missing appointments, and refusing to coordinate reasonable time windows for appointments with their wholesale customers, thus forcing their competitors' customers to lose a full day of work, just to be home to meet with the twin installation crews required to obtain the basic DSL service. If an appointment were missed, the customer faced the potential of losing a second day of work in order to have his or her DSL installed. In what could not have been a coincidence, it was not unusual for a marketing representative of the ILEC to contact the competitor's prospective DSL customer after the reseller had notified the ILEC of the impending cutover. The ILEC salesperson often sought to sell the putative reseller customer the same service being offered by the reseller, often at a lower price, with a promise that it could be installed within a narrow appointment time window on the targeted installation day.

It was also common for the new service providers to invest in automated provisioning systems to streamline their customer relationship management processes. Unfortunately, for them, the ILECs almost always demanded that manually prepared facsimile order forms be sent to initiate service orders. The ILECs made this demand despite the probability that they could have easily absorbed the capital and operating net cost of accepting automated transfers. The manual processes had several defects. They were an unnecessary expense to the competitive providers in both additional capital and additional operating expenses. They also often resulted in transcription keystroke errors, which led to additional correction costs, delayed the activation of the switchover of the proposed customer, and created an image of poor quality among the competitive DSL providers.

We do not know how pervasive these practices were, or to what extent to they were actions of rogue employees resentful of the new competitors and concerned about job security, or if they were actions sanctioned or at least condoned by senior management of one or more of the ILECs. We are, however, convinced, on the basis of our own experience and many anecdotes from others, that these practices were sufficiently widespread to have had a seriously negative impact on the DSL resellers. Were these actions just coincidence, or did they represent unfair competition? It was an argument that never got resolved, because both the DSL access providers, and most of their reseller ISP customers failed to survive to wage an effective challenge. In the same period, the ILECs were openly exercising their legal prerogatives to challenge various elements of the 1996 Telecom Act and the FCC's implementations of it. For instance, the ILECs filed lawsuits to seek redress against what they claimed were constitutionally illegal "takings" of their assets under the Act. These rearguard actions extended the resolution of many critical commercial settlement matters between the new competitors and the ILECs. It also muddied the water for numerous new competitors seeking to finance their operations and indeed to obtain fair treatment for the ILECs on reciprocal compensation and related matters. The ILECs had an unfair advantage in the legal challenges. They had orders of magnitude larger war chests, much of which was built upon monopoly utility profits. Because they had much larger cash flows and a much larger customer base, they could easily afford the cost and uncertainty of lawsuits.

The net impact of the myriad complaints lodged by upset customers against the new competitors was immense. Every truckroll required to provision a new DSL customer cost the new competitor approximately one year's profit from a single customer. The new DSL innovators and their resellers faced substantial financial obstacles quite apart from the nonavailability of capital in attempting to implement the spirit of the Telecom Act. Ultimately, virtually all of the leading firms in the DSL sector were either merged with or sold in distress to the ILECs, ending the grand experiment of a forced facilities resale concept advocated by its authors in Congress. To this observer of the "law of unintended consequences," it has been proven beyond a doubt that the only telecommunications service providers with any substantial chance of developing into true competitors to the ILECs will be those that have access to their own autonomous infrastructure or those that can operate fully open, shared public networks solely on a wholesale basis, open to all comers on equal terms, as has been successfully implemented in the United Kingdom with the structural separation of British Telecom (BT).

Reassembly of the Old Telephone Alignments

Many of the major new telecom providers only avoided bankruptcy at the beginning of this century by merging with established companies that had sufficient cash to sustain the significant operating losses that were common among carriers of all types. The Qwest merger, in reality their absorption by the former ILEC/RBOC, US West, is a prime example of the integration of the new innovative competitors into the establishment. One interesting and valuable aspect of this trend was that the legacy players were able to upgrade and expand their facilities, often at a substantially lower cost than the actual market value of the facilities, through these acquisitions of the newcomers out of or on the brink of insolvency. As a result, the innovative state-of-the art infrastructure installed by the new entrants in the late 1990s, and then purchased by the legacy carriers, now represents a significant portion of the highest quality telecommunications infrastructure in the United States.

Since the almost complete collapse of the leading organizations challenging their market dominance, the incumbent telephone companies have spent most of their energies on acquisitions and consolidation of their market positions rather than investing in upgrades or extensions of their facilities and customer base. In the United States, this activity has amounted to a reintegration of the traditional telephone industry.

As we have shown earlier, for almost a century before "divestiture" of the Bell System and GTE in 1984, the U.S. telecom industry was dominated by one vertically and horizontally integrated giant. The Bell System combined the dominant LD provider, AT&T Long Lines, and seven RBOCs, which were each the dominant ILEC in their region, besides being the leaders in technology development (Bell Labs) and in manufacturing (Western Electric). The mergers allowed by the 1996 Telecom Act over the past decade have lead to the emergence of two massive organizations, which together dwarf the old Bell System. Verizon was created by the merger of Bell Atlantic and Nynex and then added the assets of the largest CLEC, MCI-Worldcom assets, and most recently announced the acquisition of Alltel. Southwestern Bell rolled up four of the other former seven Bell "sisters" along with AT&T Long Lines. Southwestern Bell, renamed first as SBC and now as AT&T, consolidated AT&T IXC operations, Pacific Bell, Ameritech, Bell South and the AT&T Wireless operations, and Cingular Wireless.

In at least one critical aspect, these two consolidations have a wider, potentially more pernicious, impact on the consumer than the old Bell System. At the time of divestiture, the mobile telephone industry was in its infancy. Although in the United States AT&T pioneered the mobile industry, the early leaders were independent innovators such as Craig McCaw, who originally made his fortune in the cable industry and who now heads Clearwire, a leading firm in Mobile WiMAX implementation. Now, however, as we will show in the next section, the top two mobile providers, AT&T and Verizon, are also the two largest landline and cellular service providers. Thus, the reintegration of the old-line telephone industry has extended its oligopoly into the mobile arena.

Worse news awaits consumers in that AT&T and Verizon are in by far the best position to dominate the quadruple-play offerings that will become essential buys for most of them. Quadruple play refers to the provision of voice, data (primarily Internet access), television, and mobile by one carrier—the ultimate in one-stop shopping. Many years after the first promise of cross-industry competition between the cable and television industries, and some years after the industries publicly backed away from direct competition, telephone companies are now deep into the implementation of fiber-based television services to the consumer. Both AT&T and Verizon are presently touting fiber-to-the premise (FTTP) implementations to deliver fully converged video, voice, and high speed data solutions to their customers. Verizon is further along with their deployments, claiming that about 7 million homes passed circa December 2007; yet, compared with its existing almost 48 million access lines, the subscriber uptake of only about 515,000 FiOS subscribers represents only a tiny fraction of its service footprint.

The direct competition in live video offerings between the cable and television industries, coupled with the limited competition from direct-to-the-home satellite offerings, is certainly a boon for the consumer, at least at first blush. The shortcoming stems from the fact that the cable television MSOs do not now offer mobile services. Thus, the cable MSOs cannot, at present, effectively compete against the telephone industry in quadruple-play offerings. The recent announcement of the Comcast, Time Warner, and Bright House participation in the newly merged and expanded Clearwire mobile WiMAX rollout may alter this situation dramatically.

Prior to the Clearwire initiative, the leaders in the cable industry had developed a tentative plan to remedy this severe competitive handicap. Led by Comcast, the majority investor, and accompanied by Time Warner, Cox and Bright House Networks, the larger MSOs formed Spectrum Co. and paid $3 billion for 30 MHz of advanced wireless services (AWS) spectrum at 1.7 and 2.1 GHz in FCC Auction 66. This spectrum can be used for 3G and 4G mobile services. However, the timing and methodology for rolling out mobile services using this spectrum are uncertain. The cable MSOs have been largely silent on their plans, which, in any event, appear to have suffered a setback because of the withdrawal of their only experienced wireless player. In August 2007, Sprint, which had paid $100 million to be part of Spectrum Co., withdrew as part of the substantial reorganization and retrenchment caused by its losses in its basic cellular services. Although the newly acquired mobile spectrum should allow the cable MSOs to eventually compete in quadruple-play offerings, they will be lagging behind by a significant margin for some time to come. The cable industries' serious handicap in competing in quadruple-play offerings is exacerbated by their handicap in competing in the triple-play offerings. The cable industry leads the telephone industry by wide margins in its core business of television fare. However, the MSOs are latecomers in the other two triple-play offerings, voice and Internet access. The MSOs have strong offerings in digital voice and Voice over Internet Protocol (VoIP). Their market shares, however, remain much smaller than the ILECs, and as they endure the inevitable growing pains to design, implement, and shake down their installation, maintenance and repair teams, they too face the conundrum that so afflicted the CLEC industry: how to build a nationwide customer care system and resources from scratch to compete against one that was built and continually exercised over decades. Meanwhile, the ILECs have not been idle. They have been constantly improving their Internet DSL service, extending the range they are able to serve customers from the central offices by improving signal quality and bandwidth. The ILECs continue to lead the cable industry in Internet market share and, by a wide margin, in voice market share.

In sum, from one perspective, consumers are right back where they were in the early 1980s—faced with a reintegrated telephone industry. On the other hand, they have options that did not exist then—competitive triple-play options from the cable industry in a growing number of communities and the prospect of a quadruple-play option from the cable MSOs. Furthermore, the ILECs have behaved as they used to when they were a monopoly. Given the collapse of the equity base of the competitors among the CLEC and ISP communities, it seems logical that the ILECs would have invested most of their capital in their own infrastructure. They might have been able to secure a "game over" unassailable position of market dominance. Unfortunately for their shareholders, but fortunately for the consumer, the ILECs' focus on mergers and acquisitions of legacy infrastructures and a legacy customer base largely consisting of traditional telephone service customers has left them exposed to the efforts of their more innovative competitors in triple-play services. Despite the ILECs' best efforts at improving their DSL service, cable MSOs are offering a superior service. Once these cable MSOs raise their digital voice and Internet customer service to ILEC standards, they will inevitably trump the ILECs' DSL services. However, had the ILECs invested widely in the deployment of FTTP earlier, they could have achieved the virtually unassailable position of market dominance. Their failure to have implemented earlier either widespread fiber or DSL network upgrades has left them exposed to the risk of a substantial diminution in their market position, should their emergent broadband IP-based competition in the fiber, coax, and wireless domains execute their comprehensive deployment plans in a timely manner.

The Wireless Incarnation of the Telephone Companies and Their Continued Attempts at Defending Their Market Power: "Walled Gardens"

As noted earlier, at the same time as the ILECs were reconsolidating into institutions of almost unprecedented scale, they have also been rolling up the cellular telephone industry at a rapid clip. AT&T and Verizon are now the largest and second-largest cellular operators, respectively. Their combined share of the wireless industry is approximately 53% (Fig. 1.1).

The market power that their size in the mobile industry affords AT&T and Verizon has been reinforced and perpetuated by the practice of creating what are commonly termed "walled gardens." This practice is not the province only of ILECs. It has been a common practice among all mobile service providers since the genesis of the mobile industry almost three decades ago. It perpetuates the market power of the large mobile providers, while simultaneously crippling

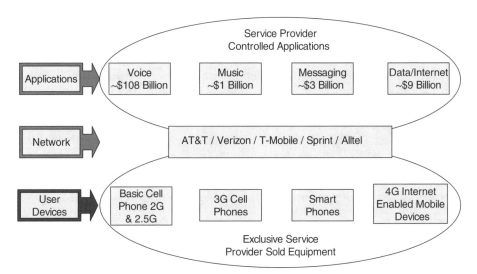

Figure 1.1. Cellular oligopoly in the United States—the "walled garden" model. Competition among the cellular operators has been limited to the horizontal plan.

innovation and competition among the vertical markets that support the cellular operators for subscriber equipment, content, and applications.

The rubric "walled gardens" refers to the practice, familiar to all consumers, whereby the mobile provider erects a barrier against the consumer using a mobile handset other than what the mobile carrier blesses. This prohibition is reinforced and extended by the common practice of selling the handsets at rates that are substantially below standard retail prices. The quid pro quo for the carrier is a long-term contract with substantial penalties for early termination. The consumer is thus locked into a particular carrier that provides a limited range of equipment for an extended period of time. It becomes very difficult for a new service provider or reseller to compete. The market becomes, as it has in this instance, much more concentrated than it would have been otherwise. The consumer suffers in multiple ways, as he or she always does when competition is stifled—less price competition and fewer innovations in service features and pricing plans.

This type of tight control over subscriber equipment and discrimination against applications and equipment not sourced through the mobile carriers are in direct opposition to the long-standing policies that were imposed on the legacy wireline telephone companies in the Carterfone decision of 1969.* The consumer and the industry would be better served if the cellular industry were to adopt a universal attachment equivalent to the ubiquitous RJ-11 at the terminal edge of the wired telephone network.

The equivalent opportunity in the wireless industry is the common air interface that has always existed under the various cellular standards, but that has been consistently blocked by carriers to limit access to phones sourced only through the carriers' own distribution channels or resellers. There is, however, late breaking news in this regard from the second-largest U.S. mobile carrier, Verizon. The final portion of the next section will address the very encouraging announcement regarding its pledge to open its network to any compatible user equipment by November 2008. On March 18, 2008. Verizon released its initial policies regarding just how "open" it was going to become. It will remain the gatekeeper for certification of all devices to be enabled on its new "open network." Thus, it will take the market a while to determine just how unfettered Verizon actually intends to operate with third parties selling devices directly to consumers for use over its network, and how applications developers will be able to deliver its services openly to individual users without cutting Verizon into the revenue potential beyond their income for providing IP access and transport.

The Pending Deployment of Mobile WiMAX Overlay Networks

The other players in the cellular industry, which lag behind the former members of the Bell System in size, have also been expanding through sequential acquisitions of operating companies and spectrum assets. Sprint Nextel, T-Mobile, and Alltel have emerged as the primary alternative players in this space. Sprint's combination

---

* The landmark Carterfone decision by the FCC, June 26, 1968.

with Nextel uniquely positions it to pioneer next-generation broadband wireless mobile services using the spectrum that it now commands post the acquisition of the MCI-Worldcom MMDS (now BRS—Broadband Radio Service) spectrum holdings and the spectrum resources that were in Nextel's possession prior to the merger. The Mobile WiMAX division of Sprint Nextel has been named XOHM. Clearwire, Craig McCaw's entry into the Mobile WiMAX market, is another major spectrum holder with the resources to potentially challenge the legacy cellular operations of the ILECs.

On July 19, 2007, XOHM and Clearwire announced, with much fanfare, the creation of a partnership for the mutual development of a nationwide Mobile WiMAX network. They proposed to divide their markets, with each concentrating on a particular region with common branding, roaming agreements, and a number of other efficient sharing arrangements. The promise was the first, and quite substantial, build out of next-generation fixed and mobile wireless facilities using the WiMAX protocols by a combination of resources fully capable of challenging ILECs' mobile supremacy.

Curiously, less than four months after announcing their intention to jointly pursue the nationwide deployment of Mobile WiMAX services, XOHM and Clearwire announced that they were unwinding their combined efforts. The change in leadership at Sprint, with its board of directors replacing Gary Foresee with Dan Hesse as its CEO, and continued challenges to the consolidation of the Nextel iDEN™ (Integrated Digital Enhanced Network from Motorola)-based network with Sprint's core mobile PCS business based on code division multiple access (CDMA) technology, apparently led to the inability of the two parties to reach a mutually acceptable agreement. Because of the long history and close association between Mr. Hesse and the Seattle wireless community, these discussions were reignited, and the subsequent merger of Sprint Xohm and Clearwire was finally announced on May 7, 2008. This new combined organization will also be sponsored by a consortium consisting of Intel, Google, Comcast, Time Warner Cable, Bright House networks, and John Stanton's Trilogy Equity Partners. At the time of this writing the merger is in the approval process by the FCC and the Justice Department with a closing expected during the fourth quarter of 2008.

Breaches in the Walled Gardens. To what extent will the new and expanded Clearwire affect the deployment of comprehensive Mobile WiMAX overlay networks? Suffice it to say, at the time of this writing, that question looms large and significant, with no answer yet. However, there was a positive aspect to the XOHM/ Clearwire merger that might have favorably influenced similar developments for Mobile WiMAX networks by other potential competitors. The Clearwire carrier combination has pledged to deliver services over an open network interface that would allow customer-owned devices to have unfettered access to the Internet, i.e., a consumer could connect any device that he or she wished to a Clearwire-provided mobile network connection, regardless of whether it was furnished or approved by Clearwire. This is the first of the anticipated wireless broadband extensions of the global Internet to any location using small mobile devices.

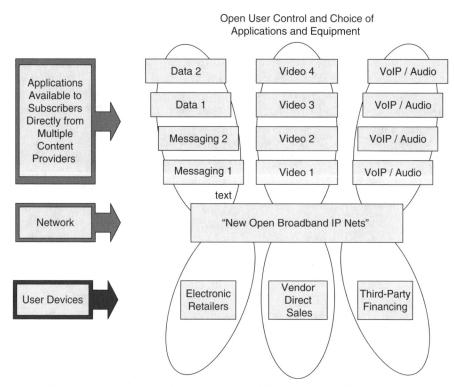

Figure 1.2. Open cellular services model—the "unwalled garden."

The Clearwire concession to open access was but the first breach in the previously sacrosanct walled garden business model that the mobile providers had been able to impose for more than two decades. Their proposal was followed soon after by Google's open and notorious open access lobbying in connection with the 700-MHz spectrum auction, which was seconded by other participants. The FCC responded favorably to this "open network" proposal by incorporating a requirement for the 700-MHz auction to mandate open access if a minimum bid of \$4.6 billion was obtained for the C Block spectrum. The hurdle was met, and both Verizon and AT&T dominated the bidding, with Google not purchasing any spectrum in the auction, but providing Google with a huge win based on regulatory lobbying, and retaining its neutral stance as a software, content, and advertising entity among all service providers. This section concludes with the fairly safe prediction that the era of the walled garden is coming to a well-deserved end, to the great benefit of the consumer and of market competition.

The next generation of the wireless broadband–enabled world will be filled with adaptability and user-controlled preferences and choices among content providers of all types, from basic voice to video interactivity (Fig. 1.2).

700-MHz Auction Participants and Results

At the time of this writing, the FCC has completed the 700-MHz auction. The public–private D Block spectrum did not receive sufficient bids to clear the reserve price. The FCC and Congress will reevaluate this failure to thrive and decide how to bring this unique public–private initiative into reality. The following is a brief discussion of the issues and proponents of these new business models.

Cyren Call. One of the most noteworthy recent FCC regulatory developments was the creation of a mandated public–private partnership. Under this approach, the winner of the D Block portion of the 700-MHz auction would have had an additional 10 MHz of spectrum added to its own in exchange for voluntarily constructing a nationwide public safety wireless broadband network that leverages the commercial network's infrastructure, backbone, and applications. Morgan O'Brian, one of the original founders of Nextlink, has promoted this concept. The FCC adopted rules to authorize this scheme in July 2007. The D Block commercial licensee would gain access to use the spectrum licensed to the Public Safety spectrum to provide nonpriority wireless broadband services to commercial subscribers, while simultaneously meeting Public Safety's critical communications needs in key areas such as network coverage, availability, and reliability and ensuring that Public Safety users will automatically receive first priority access rights on the network.

On October 5, 2007, the Public Safety Spectrum Trust Corporation (PSST) announced the appointment of Cyren Call Communications Corporation as its advisor. This confirms Cyren Call as the primary liaison between the PSST and the commercial sector, including all parties interested in bidding for the upper 700-MHz D Block license that will seek to partner with Public Safety in the creation of the nationwide, shared use wireless broadband network.

Frontline Wireless. Frontline Wireless was organized by a team of telecommunications industry veterans with deep roots in the government and in industry. It had expected to be a major bidder in pursuit of the 700-MHz D Block spectrums. It had assembled a team of luminary participants, sponsors, investors, advisors, and lobbyists to support its initiative. Unfortunately, the organization announced its nonparticipation in the auction at the deadline for making the bidding credit deposits.

As originally stated on the Frontline Wireless website, "Frontline Wireless envisions a 4G wireless broadband network that will make advanced Internet services as ubiquitous as the air we breathe. By leveraging efficiencies of shared spectrum and network infrastructure, Frontline will empower first responders with state-of-the-art technology and liberate consumers from the 'walled gardens' of the incumbent wireless providers."

In addition to meeting public safety agencies' needs for fully interoperable broadband communications networks, Frontline had also promised to provide fully open access services to commercial service providers and the general public.

Had Frontline been able to deliver on its promises, it would have created a formidable counter to the legacy cellular operators, with the implementation of a true nationwide 4G network well in advance of the technical evolution Cellcos hoped for. There have recently been calls for investigation into the reasons for the collapse of Frontline and the events that led to its sudden demise.

FLEXIBILITY COMES TO WIRELESS SPECTRUM

The Demise of Regulatory-Designed Single-Purpose Spectrum

Almost since the advent of radio, different wireless services have been provided over discrete networks and discrete portions of the radio spectrum. We obtained our radio services over separate slices of AM and FM radio spectrums. Our television was broadcast to receivers using separate portions of the VHF and UHF spectrum. Our cellular phones were brought to market using the 800-MHz spectrum and 1800-MHz spectrum bands, which were allocated by the FCC for the provision of what were initially just voice communications. Virtually all of the radio spectrum was sliced and diced into small portions, which were designated by service rules for specific applications and services.

The lengthy process by which the FCC, circumscribed by International Telecommunications Rules, adopted these narrowly defined service rules limited innovation and forestalled the introduction of new services. The incumbents in a particular service used the rule-making process to devastating effectiveness in protecting their domains and investments. Often, by the time the service rule-making procedure had been completed, technological advances, for instance, in dynamic bandwidth reallocation and merging of fixed and mobile uses, rendered the rigid rules obsolete. Michael Powell, former chairman of the FCC, stated the issues succinctly at his press conference on "Digital Broadband Migration" on October 23, 2001. Regarding spectrum allocation policies, he stated,

> Put simply, our Nation's approach to spectrum allocation is seriously fractured. There have been dramatic changes in spectrum requirements and technology and services that use spectrum since 1934. Yet, while we have made some major strides in how we assign spectrum (principally through auctions), allocation policy is not keeping pace with the relentless spectrum demands. The spectrum allocation system is not effectively moving spectrum to its highest and best use in a timely manner.
>
>   The central problem with our current approach is that it is a command and control approach that requires government officials to determine the best use for spectrum and to constantly change the allocation table to accommodate new spectrum needs and new services. This is becoming an impossible task in today's dynamic environment.
>
>   The consequence of our current system is that it is entirely reactive. With new emerging uses, the Commission must not only evaluate and react to the new services, it must also deal with the conflicting set of legacy allocation decisions. New services are forced to demonstrate demand for the service to

justify modification of the allocation table. Lack of proof, however, makes it hard to do so and unleashes a highly politicized process. Existing users move to block new uses and line up support for their position, and the new providers are forced to do the same. The ultimate decision is reached as a result of a politicized reactive process.

Additionally, spectrum allocation policy provides few incentives for using spectrum efficiently. Existing holders have little incentive to consider using their spectrum for more valuable uses, since allocation restrictions will prevent consideration of alternative uses. Moreover, once a carrier obtains spectrum it has little incentive to use it efficiently where there is no flexibility.

Any policy change must respect that spectrum is a public resource and must be employed for the benefit of consumers.

Over the past decade, the FCC has been slowly but inexorably relaxing its restrictions. More and more frequently, the FCC is allowing the licensee, particularly in the commercial transport services addressed here, to provide any type of service it wishes, subject only to its noninterference with other uses within or adjacent to the band.

The advent of the wireless mobile broadband Internet will accelerate the trend to abandon these traditional, narrowly defined approaches to spectrum management and radio services. Generic wireless broadband IP networks will not require spectrum partitioning and isolation. Instead, multiple applications and services will share network resources. Network operators will isolate their discrete information through the creation of virtual private networks and security measures to maintain privacy and the security of the data streams. Discrete vertical market applications will be driven by software, not by dedicated narrowband spectrum allocations and restrictive service rules.

THE WIRELESS TECHNOLOGY DIASPORA

There has been an almost complete shift of telecommunications equipment manufacturing to developing nations and third world countries. The major infrastructure and user terminal manufacturers have evolved their business models to leverage the outsourcing of the manufacturing process. Silicon Valley vendors have become essentially "virtual manufacturers," benefiting via cost reductions through outsourcing many development and manufacturing tasks (no permanent labor force) and in being able to shift product lines quickly to adapt to rapidly changing market requirements.

As the capability to develop and manufacture complex wireless infrastructure hardware has extended to the developing world, industrial development policies have emerged to provide both financial incentives to carriers and various protections to domestic national interests and franchises. Early examples of this trend were evidenced in Japan, which used its domestic regulatory process to localize technology standards and products that were primarily developed and supplied by its local vendors. Korea followed the Japanese model for industrial development, and most recently, we have seen China pursuing a similar course to ensure that a significant portion of its telecommunications infrastructure is sourced from its domestic industry. This local specialization has often led to the creation of local subsidiaries by multi-national manufacturers within China to address local market opportunities in their burgeoning markets.

Specific examples of these industrial policies include the creation of the WiBro (Wireless Broadband) standard in Korea, the adoption of domestic standards for 3G services (TD-SCDMA) in China, and the historical barriers erected around the European Union (EU) through various European Telecommunications Institute (ETSI) standards, which are not seamlessly interoperable with many open standards. In addition to local standards, control is also exercised through the application of protectionist policies in the allocation of radio spectrum for the benefit of local operators and equipment manufacturers. In addition to policies are the often attractive financing terms that governments extend to their domestic vendors to allow them to compete effectively and powerfully in the growing international market. Thus, companies such as Huawei Technologies Co. Ltd. (Huawei) and ZTE are capturing significant market share internationally, in large measure on the basis of their lower costs and the ability to offer attractive financing terms for their customers.

The refinements and advancements that have evolved from the combined contributions of each of these major sector development groups over the past decades have often been constrained by regulatory delays, patent and standards disputes, and the highly volatile state of the capital markets. Like so many technology-driven sectors of the economy, wireless system innovators have often faced the disruptive effects of financial market support, swinging wildly between buying frenzies and intense sell-offs. The names of the promoters and services may have changed, but the challenges and issues facing the wireless industry are clearly repeating the recurring themes from our twentieth-century history of innovation.

As we rapidly move further into the twenty-first century, our pace of development and change is accelerating, and the macro trends that will affect our industry are coming into focus. The subsequent chapters will attempt to define the key elements that will both drive us forward and slow us down as an engine of change that has proven itself capable of creating true revolutionary change throughout society in all regions of the world.

The net result of this global dispersal of wireless technology and development will exert an ever-increasing pressure on the science and innovation drivers of the industry. It is critical to note that the early work on commercializing OFDMA technology, a key technology in 4G wireless systems, was led by Chinese and Korean nationals. Some of these innovators had worked in the United States in early stage technology start-ups, while others were with large government research institutes and large OEM electronics–manufacturing firms. Regardless of where they were, the inventor's listed on the core patents for next generation OFDMA wireless broadband systems are predominantly Asians.

The unique contributions of brilliant individuals continue, and the authors predict that some additional names will soon be recognized among the pantheon of wireless technology innovators. The authors alert the readers to watch for the emergence of Dr. Hui Liu, Dr. Alex Li, Wenzhong Hong, Wei Tang, and several other innovators whose efforts have led directly to recent breakthrough developments in OFDMA technology. This team of technology developers worked together to develop and patent much of what is now the core technology being embraced by the global wireless industry for OFDMA. This core technology has been incorporated in the new Mobile WiMAX standard (IEEE 802.16[e]) and has also captured center stage for inclusion in the 4G standards for cellular technology. Although the WiMAX Forum and the cellular industry's 3GPP organization LTE are presently divergent in their 4G technology roadmaps, they both embrace OFDMA as the core wireless technology for the next generation of mobile wireless systems. The merits of OFDMA technology are discussed in detail in Chapter 10.

Recently, a number of early stage firms have found both investment and market traction in emerging markets, including India, China, and Eastern Europe. This development is in sharp contrast to similar early stage ventures in the United States, which have been struggling through the challenges left over from the recent depression in the sector and whose ranks have largely been diminished through their sale or merger with larger, more established concerns. The returns that they delivered to their investors were far below the range that VC investors require to sustain continued interest with investment in the sector. Much excitement was seen at the time of the Qualcomm acquisition of Flarion for about $600 million, but in reality, this represented only about a 3× return on the invested capital over almost eight years. Recently the acquisition of Navini by Cisco generated a purchase price that returned only about 2× to the venture investors over nine years. Earlier, IPWireless was sold to Nextwave for a price that was substantially less than the invested capital. The failure to deliver financial returns that create enthusiasm among the VC investment community will negatively affect the next wave of wireless innovators in the United States. When market conditions and risk capital are readily available to innovators in emerging nations, it will result in a continued shift of technology development and new venture formation to more friendly locations. The migration from long-term patient investment in core technologies to shorter-term, higher-return investments will lead the U.S. VC industry to squander its attention and capital in the current bubble market for Web 2.0 Social Networking ventures. The resulting scarcity of risk capital in the United States to fund new wireless technology ventures will contribute indirectly to the continuing international diaspora of wireless broadband talent and opportunities.

The message is simple: The United States is at immediate peril of slipping from its position of global technical leadership in the wireless broadband industry into a role of just being a "trading nation," relying on profits of being a financial intermediary rather than being a true value creator among the leading technology development and applied technology commercialization nations. There is an urgent need in the United States for a comprehensive industrial development policy, tied closely with increased attention paid to our growing technical education gaps with competing nations worldwide.

CELLULAR CARRIERS: STUCK ON STICKINESS

Our almost universal adoption of mobile phone services has conditioned our society to expect certain things about their service and to accept (through lack of any substantive competition to date) an arbitrary set of rules that define our commercial relationship with our service providers. Legacy cellular operators have created numerous marketing programs, with policies designed to maintain the customer relationship (the "stickiness" factor). Long-term contracts tied to free or discounted phones are the norm. The introduction of number portability a few years ago broadly exposed the "fine print" requirement for customers required to buy out the full-term value of their cellular contracts before being able to port their cellular number to another carrier. This industry-wide sales policy has created pretty extreme stickiness, by almost any measure. The impact on wireless customers has been significant, and given the high degree of vertical integration that has resulted from the consolidation of the leading wireless service providers back into the largest wireline telephone companies, it will be extremely unlikely that these market leaders will lead with innovation.

Verizon and AT&T now dominate the industry post their reconsolidation, and these firms have demonstrated the most onerous business practices in the quest to keep customers captive to their services. Locking phones and imposing network control to keep phones not obtained from the carrier off the network are their standard business practices. Further, their data services have contractual limitations on the use of bandwidth that rigidly constrain subscriber use of their Internet access services. Although these data contracts have apparently been crafted to prevent abuse by subscribers using their IP connections for applications that far exceed what the carriers have determined are "normal usage limits," they also prohibit the use of these systems to access solutions that may compete with their core voice and walled garden data applications, such as VoIP services that use software applications like Skype, Gizmo, or other "soft phone" clients, which could easily be integrated into cellular handsets. Franchise protection and customer retention have become a high priority among U.S. and many international cellular operators, with a wide range of defensive strategies being tested or implemented, including adding packet time delay (degrades VoIP service quality) and network sniffing of unauthorized applications. Protectionism is growing in proportion to the increase in the power and availability of Internet-centric, open source data services, spoofing rigidly controlled legacy-closed service bundles.

"Unsticking" the Carrier Relationships

As we move toward the inevitability of open wireless broadband network availability, the marketing crossroads will be where the Internet culture of open access and open systems collide with the traditional telephone system–inspired closed architecture of the cellular operators. The authors predict that there will be a wide range of new business models trialed by new service providers, ranging from traditional subscription services to permutations of prepaid, pay-per-consumption, metered services with postpaid billing, and at the other end of the marketing spectrum, free, advertising-supported wireless broadband services can also be expected to emerge from some service providers.

The next wave of wireless competition has already begun on the heels of the Clearwire Mobile WIMAX deployments. Clearwire has claimed that they will pioneer open access to their broadband IP network to allow any Mobile WiMAX–certified compatible user devices conforming to the 802.16(e) standard. If their large-scale network deployments make it through the current financial challenges facing the consortium of pioneering firms, it will be the first mass-market assault on the traditional, closed approach institutionalized by the existing cellular operators that leverages crippled user terminals that are only usable on a specific network. The bifurcation of the mobile wireless industry into opposing camps with extremely different network assets and infrastructure will define the early battle lines over who will be the winners and losers among the new broadband wireless service providers.

In addition to expanding competition from Mobile WiMAX and metropolitan area–wide WiFi initiatives, there will soon be other new competitors emerging from the 700-MHz auctions and from among other large-scale spectrum holders such as Nextwave. Although the authors believe it will take the better part of the coming decade to allow sufficient time to enable the widespread ubiquity of wireless mobile broadband to be achieved, it will only take a few years to allow these new competitors to exert their influence on the market. For the first time, legacy cellular operators will face unprecedented competition that will be differentiated not just on price, but on a variety of enhanced services as well, which we anticipate will lead to their bringing LTE service upgrades to market sooner than if there was no new competition. The migration of the voice-centric cellular networks to broadband data services will inevitably be accelerated when carriers are faced with substantive market competition from new IP data–centric wireless operators. It is important to note that the legacy cellular operators have been struggling to keep up with the demand for increased data services and a burgeoning industry-wide belief in the extension of broadcast and on-demand video services into the personal mobility market.

Neither the existing GSM nor CDMA networks are particularly well suited to deliver broadband IP services to their subscribers. The U.S.-based GSM carriers T-Mobile and AT&T (including Cingular) have been slow to widely deploy WCDMA in the United States, which will require 5-MHZ-wide channels to implement. To date the GSM carriers have relied heavily on GPRS* and EDGE† to deliver most of their data services, which co-exist within the standard voice channel framework of 200-kHz-wide radio frequency channels with eight TDMA (time division multiple access) time slots. The economics of GPRS are

---

*GPRS—General Packet Radio Services. GPRS is a packet-based wireless protocol for integration with 2.5G GSM networks featuring data rates from 56 up to 114 Kbps.

† EDGE—Enhanced Data GSM Environment. EDGE is a faster version of GPRS wireless data service. It enables data to be delivered at rates of up to 384 Kbps.

extremely painful to the voice revenue capacity and network efficiency for GSM operators. GPRS typically bonds four (of the 8 available) TDMA voice quality time slots to deliver what is in reality only performance at the low-end of wideband speed data service. T-Mobile has responded aggressively to this challenge by pioneering WiFi hot spot access to augment its cellular data services, and recently to pioneer Unlicensed Mobile Access (UMA) services with the addition of dual mode GSM/WiFi phones and home installation of T-Mobile hot spots.

U.S.-based CDMA carriers, dominated by Verizon and Sprint Nextel, have fared marginally better with their delivery of data services using evolution–data optimized (EV-DO) network enhancements and wider deployments of CDMA2000 3G infrastructure in numerous markets. These data-centric network elements in heavily loaded networks are still only capable of delivering wideband services to large numbers of simultaneous users, which is substantially less than one megabit per second. However, it should be noted that the implementation of data services is consistently managed on discrete channels that are not shared with voice services. Indeed the incorporation of 3G into the CDMA2000 network operators has in large measure been consumed by increased voice traffic, rather than portioned to support significant data services.

As mobile wireless systems continually strive to keep abreast of the burgeoning customer demand for both basic and enhanced data services and high speed Internet access, the need to be as spectrally efficient as possible will emerge as a key determinant of network technology selection, as the subscriber base has now grown to over 200 million in the United States. As we look back at the digital evolution of the cellular industry, we can observe this trend in retrospect. The original cellular networks were based on FM analog radio technology, and required a discrete 30-kHz radio frequency (RF) channel to support each voice conversation. As we moved forward with the original migration to digital technology TDMA techniques were adopted (see IS-54), supporting four simultaneous voice conversations within each 30-kHz RF channel—a fourfold increase in efficiency over the analog systems. In parallel, the EU also adopted TDMA technology as the core for the GSM standard, which was enabled using 200-kHz RF channels with eight time slots for carrying traffic. Subsequently, Qualcomm began advocating CDMA as a more spectrally efficient means of delivering voice services and drove through the second U.S. cellular standard (IS-95). The CDMA implementation used 1.25-MHz wide RF channels that typically support 64 simultaneous voice conversations. In practice, the CDMA solutions are between four and eight times more spectrally efficient than the TDMA systems. The growing scarcity of radio spectrum in the frequencies that are practical for mobile non-line-of-sight services makes spectral efficiency a central issue for any new mobile system architectures to consider. This requirement is now driving all wireless network planners to seek the next level of spectral efficiency that has been demonstrated by OFDMA technology, and incorporated in the 802.16(e) standard for what is now being commercialized as Mobile WiMAX.

Both the cellular industry's 3GPP technology organization and the WiMAX Forum have selected OFDMA as the technology of first choice for the 4G mobile wireless networks. Thus, we see a collision of standards derived from different roots, but with similar goals and objectives for network efficiency and performance.

The mobile wireless industry is following a bifurcated trail that leads to the implementation of 4G platforms. The existing cellular industry is evolving its 2G "narrowband" GSM and CDMA networks along the 3GPP roadmap through the 3G "wideband" era into the LTE technology, which will ultimately incorporate an OFDMA-based solution. In parallel, the emerging Mobile WiMAX carriers are already deploying OFDMA-based networks following the 802.16(e) Mobile WiMAX standard. The power of the legacy cellular operators to maintain their market dominance will thus be severely challenged as new service providers enter the market with solutions that are already as powerful and flexible as the LTE 4G vision, which is not anticipated to enter the global wireless infrastructure market until 2010. As with all things in the domain of technology in the early twenty-first century, we find the speed of change accelerating with little consideration during this change to the financial exposures of last-generation platform-based businesses. In addition, as we enter into the "long tail"* Internet economy, with mass specialization, featuring a steady decrease in horizontal homogeneous markets in favor of a very large number of specialized niche markets, the need for legacy service providers to reinvent their business models will be an essential requirement for survival. The question for investors, regulators, vendors and consumers is a big one: Can they do it? If not, how they milk their installed base and capture capital for returns in a declining market that will provide the material for a future generation of Harvard Business School case studies. Balancing survival in the midst of rapid obsolescence and technology evolution will take tremendous creativity, commitment, and investors with an awareness of the long-term benefits of supporting the wireless industry from peak-to-peak of successive waves of progress, but on a timescale that must be measured by half-decade cycles, not recurring quarterly performance panic attacks.

The business models for the legacy cellular operators, which have largely been closed proprietary environments, are also going to face competition from a number of new access models ranging from free, advertising-supported services, to pay-per-consumption metered services, and even various permutations of traditional subscription services. The U.S. GSM operators have typically constrained handsets to those purchased directly from the carrier by locking the SIM card instruction sets to limit the use of the phone to only their network—a limitation, by the way, that can be overcome by just about any independent cellular retail outlet in GSM-served nations for about $20. Similarly, the CDMA network operators have constrained independently supplied phones by blocking non-preregistered electronic serial numbers (ESN) in their switching systems.

---

* Anderson authored a book on the subject, *The Long Tail: Why the Future of Business is Selling Less of More* (2006).—Citation and Figure 3-7 from Wikipedia.

Regardless of how future users of converged broadband services obtain their connectivity, the applications that evolve to deliver the greatest personal value and utility to users of all types, ranging from personal services to commercial and government applications, will undoubtedly benefit from the availability of broadband connectivity into all application domains. Spectrally efficient broadband wireless systems will not only increase the number of simultaneous voice conversations a given amount of bandwidth can deliver, it will also allow for enhancements in voice quality and applications. Broadband wireless will also allow for the true convergence of voice, data and video, enabling all services to be available at any location from which the user may seek access.

MANAGED NETWORK SERVICES: THE OUTSOURCED NETWORK

We should not get too far afield while contemplating the cellular business of the future without examining some of the new business models that are emerging in the industry. Among the most innovative and substantive changes to the core business model of successful wireless carriers is the concept of outsourcing the entire access and transport network to third parties. The utilization of all or partial outsourcing of the network essentially transforms the core business into a pure marketing and sales organization, with a fixed cost associated with network operations for the first time. In what circumstances does this fairly radical approach make sense, and at what level of critical mass can or should it be justified? The business decisions attendant to this board room and management decision is nontrivial, and the potential for disaster is significant. However, if successful, the approach appears to deliver a very powerful transformative influence on the early pioneering operators using the outsourced network approach.

The concept for network outsourcing follows the trend in many industries to seek ways to hand off processes and business elements to specialist third parties whose concentration and scale allow them to accomplish the required mission at a lower cost, with higher efficiency, and with better quality. Responsiveness to outages and network management are efficiently removed from being a variable cost to a contractually defined set of requirements at a predetermined fixed price. If these efficiencies are indeed integrated into the outsource network services contract, then the underlying cellular service provider can concentrate management and financial resources on improving their marketing and sales efforts. Maintaining quality of service and formulas for adding capacity and new features and service platforms to the network must be carefully anticipated, and the agreements must be flexible enough to allow both parties to adapt to rapidly changing market requirements in a win-win environment. Alternatively, as soon as the balance tips into asymmetric suffering or squeezed margins, the outsource relationship will start to unravel quickly.

Typically managed network services are provided by the major infrastructure equipment manufacturers who are seeking to improve their participation in the value chain by moving deeper into services. Their unique capabilities to engineer, project manage, design, install and maintain network facilities empower these organizations to capture a position of leadership in the field. The other potential groups of organizations who are well positioned to participate in this new model are the major systems integrators (SI).

Early wireless infrastructure vendors pioneering this new model include Ericsson, with complete network responsibility for the company 3 Networks in the United Kingdom and over 30 carrier clients worldwide, and Nokia Siemens Networks, which has contracted managed services with 34 clients in 28 countries, providing managed network services for over 20 operators internationally.

One of the most comprehensive and successful users of outsourced network operations is, Bharti Airtel Limited (Bharti) in India. Bharti has contracted with a number of outsource specialists for various elements of its business operations, including IBM for IT support, Nortel for call center services, Ericsson for the management of more than 70% of its GSM network infrastructure, and Nokia Siemens Networks for the expansion of rural cellular and fixed lines. As Bharti expands into international markets, it is also using managed network services to efficiently enter new markets. For example, in Sri Lanka, its local subsidiary Bharti Airtel Lanka Private Limited, has a managed network services relationship with Huawei, for the construction and operation of a 2G and a 3G network.

A recent In-Stat [4] report has estimated the market for managed network services. The following is a summary of its key findings:

- The total mobile managed services market is expected to double over the next five years; growing from $22.2 billion in 2006 to $52.2 billion in 2010.
- The evolution of network operator infrastructure technology is creating opportunities for vendors to provide managed services to mobile service providers.
- Markets for traditional managed services, such as customer care and billing and network management, will continue to grow, but will be eclipsed by rapidly growing markets for managed network services running the entire access and transport network.
- Virtually all leading mobile infrastructure and applications vendors are developing and marketing comprehensive managed services solutions for carriers as they evolve their business models to embrace outsourced solutions.

ENHANCED BROADBAND VOICE

Even mundane voice services will be favorably impacted by the advent of broadband wireless as the value of improved audio quality is added to the delivery of traditional voice services, thus enabling a wide range of innovative new categories of voice and audio communications. High fidelity voice, stereo voice,

voice storage, audio blogs, enhanced voice mail, voice integrated into gaming, and audio text services will all benefit from the incorporation of broadband connectivity into the realm of voice communications. In addition to the increased bandwidth, the evolution of voice services have also been affected by the migration to VoIP, which has simplified the management of voice distribution to include one-to-many conference calling, group calls established on demand, push-to-talk features, and incorporating virtually any number of simultaneous participants as desired. Further, when voice is translated into an IP-compatible format, the archiving of conversations for later retrieval, and the seamless integration of voice into unified messaging systems will spawn further new applications and services that have voice as a key element.

The emerging broadband access capabilities will also have disruptive impacts in the domain of LD voice services. The cellular industry pioneered the "postalization" of domestic LD services when it combined airtime and LD services into a fused flat rate per minute. Previously, LD services were sold on a distance-sensitive basis, and the new concept treated LD the same as the postal service treated first-class mail, one price to any U.S. location. AT&T Cellular pioneered this market-changing breakthrough under Dan Hesse's leadership, with its innovation of "bucket" plans for monthly pools of airtime, which were billed on a flat rate for various amounts of service, including LD charges. Additional airtime is charged at much higher rates, incentivizing customers to purchase the largest size bucket plan per month that matches their typical usage profiles. This take or pay approach revolutionized the cash flow metrics of the cellular industry and has become the norm for virtually all U.S. cellular operators. Now the forces of the Internet are coming to bear upon both the cellular industry approach to bundled airtime and LD services by converting voice to a simple data application.

Similarly, the integration of speech recognition technologies is already making inroads into the VoIP market with new services and applications for the conversion of speech to text and text to speech. These additions to the applications developers' tool kits will inevitably impact the future of messaging systems and cross-platform digital voice.

The negative impact of VoIP on the economics of traditional cellular business voice models is potentially devastating. Presently, voice still accounts for cellular revenues in excess of 85%. The bundling of air time and LD services created a new paradigm for charging for voice services, but with the resulting fees still averaging $0.10 per minute within the presubscribed "bucket plan allocations," and $0.25 per minute if the users consume airtime beyond their subscription limits, the cellular industry remains a hugely profitable enterprise. Recently there has been a new marketing approach to bring "all you can eat" plans to the market by the cellular service providers. It is too soon to tell what long-term impact these plans will have on the industry. They are a bargain for the power users, and perhaps a means of increasing the ARPU for less voluminous users. These existing retail price plans are on a collision course with the rapidly maturing and improving Internet-based voice service delivery models such as Skype or Gizmo, wherein peer-to-peer connections are essentially free. As broadband wireless IP services become widely available, how service providers manage the competitive challenge of these alternative bypass technologies using their access networks to compete against the carriers own voice services will become the fulcrum on which the future of the industry is determined.

As voice is converted to an integrated data application element within generic IP data streams, it becomes increasingly clear that maintaining the present level of revenue from traditional voice services is highly unlikely. How the legacy cellular operators will adapt to the new realities will be a major determinant of how the wireless industry evolves to embrace the new technologies of broadband wireless, or how they will simply devolve into lesser lights of the future wireless broadband landscape. There are many parallel issues between the present market dominance and financial power of these legacy operators and their landline telephone company parents. Large top line expenses, driving business through regulatory protectionism, and supporting profitable mass market generic revenue streams, which are unlikely to be sustainable into a future populated by computing-centric broadband wireless networks. Will the legacy operators engage in the fight to maintain their market dominance through direct engagement with the new IP-centric wireless operators, or will they seek to mount a rearguard defensive posture reminiscent of that brought to bear against the CLECs and DSL service providers by the ILECs? Given that the ILECs are now the largest cellular operators, the question is whether the tiger can indeed change its stripes. Because of the self-sufficient capabilities of end-to-end autonomous networks that are anticipated in the Mobile WiMAX market, the expected collision will be one of the best wireless developments yet for consumers and for all related businesses that constitute the wireless broadband ecosystem worldwide to test these issues.

FIXED MOBILE CONVERGENCE

There are presently a number of fixed mobile convergence initiatives under way. Although one would expect that the wireline cellular operators would lead in this space, it is interesting to note the progress being made by T-Mobile with the introduction of a UMA GSM/WiFi–enabled phone package that leverages a combination of "bring your own broadband" (FTTX, cable modem, DSL, or wireless broadband) connections for installation of a home WiFi hot spot that allows subscribers to use their T-Mobile phones in a VoIP no-airtime-charge (fixed monthly fee) modality when within range of their home's or any T-Mobile hot spot. The willingness of telco-owned cellular operators to cannibalize their own revenue streams to provide competitive and innovative new services will be a large determinate of identifying the future survivors in the emerging, hypercompetitive wireless broadband marketplace. Careful and incremental management of evolutionary change is the proven specialty of large-scale telecommunications operators. The test of how well these skills will serve their pending battles with new types and kinds of competitors overlapping their legacy franchises will begin in earnest by the end of 2008. It may take five more years for the competitive landscape to mature to the point of truly competitive market engagement, but with the drive by the computing and broadcast industries into the domain of wireless personal communications, the clash of innovation versus protection of the established businesses is inevitable.

Fixed Mobile Convergence in the Enterprise

There is another emerging market that is leveraging the potential for fixed mobile convergence. Large business down to and including SMEs are being addressed by a growing number of new types of mobile virtual network operators (MVNO) whose mission is to capture the largely under-managed and out-of-control commercial cellular use by employees using cell phones in their business. The concept leverages on-site microcells that capture traffic from standard mobile handsets coordinated under the company's unified management control, typically outsourced to the system integrator or enterprise-focused MVNO service provider. Thus when employees are at work or in corporate-owned facilities, all their cellular traffic is carried via the company's internal virtual private network (VPN), and when the employees are out in the field, they are simply "roaming" onto the MVNO-underlying nationwide network, but all billing and customer support is managed by the company or its enterprise MVNO. Further enhancements to this model include the addition of PBX desk set and VoIP-based features and capabilities into the cellular handset when it is on the enterprise intranet.

Pioneers in this space include companies such as Strata8 Networks, Inc. and Sotto Wireless, Inc., both coincidentally located in Bellevue, Washington, and Spring Mobil headquartered in Stockholm, Sweden.

We anticipate that the market for managed network services and outsourced VoIP PBX services for the enterprise will evolve to become fully complementary in addressing the market requirements for fully converged services, including the mobile extension to the desktop.

BOUNDARY BLURRING

The wireline, cellular, cable television, direct broadcast satellite (DBS) television, Internet service providers, radio and television broadcasters, and content owners are all in a mad race to invade and capture enduring relationships with each other's customers. How content is made available to customers is heading for a "jump ball" environment between and among these historically dominant players in each of their respective domains. How the concept of "subscriber" evolves to address an environment of unprecedented consumer choice, wherein "customers" make content and service decisions on the basis of a comparison of all aspects of price, performance, convenience, service quality, and ease of use, will be determined in the new, complex market into which the wireless broadband industry is rapidly morphing. Responding to this changing environment is the massive opportunity facing our industry. New fortunes will be made, and others diminished, as the winners are identified among all the participants on the value chain.

The concept of convergence is finally, after a decade-and-a-half of promising talk becoming reality. Further, the concept has evolved enough to go well beyond the idea of voice, data, and video convergence to also include all aspects of mobility, leveraging the networks serving the home, the office, and mobile, delivering on-demand content to every type of platform. The ability for content providers to simultaneously deliver their products to all three screens for image and video content, including the television, the desk top or laptop computer screen, and the small-scale screen of handheld mobile devices constitutes the new fully converged quadruple-play environment.

REFERENCES

1. Juniper Research. Mobile Broadband Markets, WiMAX, EV-DO, HSPA & Beyond, 2007–2012. Hampshire (UK): Juniper Research; Aug. 2007.
2. Charles Lindblom. The science of muddling through. Public Admin Rev 1959; 19: 79–88.
3. United States District Court for the District of Columbia, Civil Action No. 82-0192, filed August 4, 1982. Modification of Final Judgment with American Telephone and Telegraph.
4. In-Stat Report. Managed Network Services for Mobile Operators. In-Stat; Copyright © 2007.

2

BROADBAND AND THE INFORMATION SOCIETY

We live in an information society and in a knowledge economy. A nation's competitiveness is directly dependent on its information infrastructure, which includes access to and availability of broadband technologies at a low price as well as access to low priced devices that drive mass adoption and usage. Availability of broadband decreases the digital divide and allows consumers to access at a much faster rate, changes user behavior, has a positive impact on the local, regional, and national economy, and in general is beneficial to the overall knowledge society. Broadband networks reduce the disadvantages of low population densities in rural areas. New applications such as telemedicine, e-government, public safety, e-commerce, small business assistance, and entertainment can be launched and adopted. As a result, new jobs are created and often new or offshoot industries. The economic benefits of broadband can also be attributed to indirect factors, including "increased commerce, reduction in commuting, increased consumption of entertainment, and savings in health care as a result of sophisticated tele-medicine" [1]. In the United States, several studies have been released that detail the prevailing economic benefits of broadband deployment.*

---

* Examples of such studies: Robert Crandall and Charles Jackson, "The $500 Billion Opportunity: The Potential Economic Benefit of Widespread Diffusion of Broadband Internet Access 2001," "Broadband Bringing Home the Bits," National Academy Press, 2002. In addition, there have been several local and regional studies looking at the impact of broadband to their economies such as "George Ford and Thomas Koutsky, Broadband and Economic Development: A Municipal Case Study from Florida," 2005.

Wireless Broadband. By Vern Fotheringham and Chetan Sharma
Copyright © 2008 the Institute of Electrical and Electronics Engineering, Inc.

There is general agreement that information and communications technologies (ICTs) have a positive impact on the broader economy by improving productivity of workers and making consumers more informed. U.S. enterprise customers, especially mobile workers, increasingly rely on high-bandwidth capability of aircards and handsets to access and deliver information. Consumers also rely on broadband capability of their devices to interact, communicate, share, publish, and be creative. Enterprises rely on broadband capabilities to send and receive emails with large attachments; browse Internet, intranet, and extranets, access corporate applications such as Customer Resource Management (CRM), SCM (Supply Chain Management), Logistics, Enterprise Resource Planning (ERP), FFA (Field Force Automation), Sales Force Automation (SFA), and many others. Inability to use these applications, some of which require broadband access, will be detrimental to productivity and impact several small, medium, and large enterprises in the United States. The above applications are across multiple business segments, including, but not limited to, agriculture, communications, government including federal, state and local, field operations, construction, finance services, health care and medical, hospitality, legal, manufacturing, public safety, real estate, retail, utilities, warehousing, and banking (Fig. 2.1).

Telecommunications plays a critical role in ICT. Though both wireline and wireless have an important role because of the wide availability of wideband wireless technologies such as EV-DO and wideband CDMA (WCDMA), and the fact that mobility is a key component of any communications strategy now, the role and contribution of wireless broadband to ICT and in turn to the wider economy is quite significant. The economic performance of Japan and Korea has been correlated with high levels of investment in their infrastructure, specifically broadband. Korea and Japan have invested heavily into their wireless broadband infrastructure. In the United States, the adoption of wired broadband into the mass market will drive demand to increase the availability of true broadband services in the mobile environment.

The mobile boom over the past decade has not only created new jobs but also contributed to economic growth by widening markets, creating better information

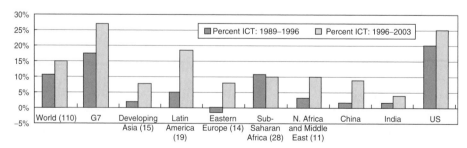

Figure 2.1. ICT capital's contribution to economic growth. *Source*: World Telecommunication/ICT Development Report 2006—Measuring ICT for Social and Economic Development, ITU, March 2006.

flow, lowering transaction costs, and substituting for costly physical transport. Apart from the impact of the mobile sector, the transformation of economic relationships and processes is particularly visible on a large scale in those countries and areas that have the highest Internet penetration levels.

IMPACT OF TELECOMMUNICATIONS ON THE ECONOMY

Telecommunication is a key component of ICTs and its impact on the economy and GDP is well studied and documented by several local, regional, national, and international bodies. In a study commissioned for the U.S. market by the Cellular Telecommunications Industry Association (CTIA) in 2005 [2], Ovum and Indepen concluded the following:

- About 3.6 million jobs were directly or indirectly dependent on the U.S. wireless telecommunications industry.
- The industry generated $118 billion in revenues and contributed $92 billion to the U.S. GDP (This number rose to $174.7 billion in 2005 and is expected to grow to $265 billion by 2009) [3].
- The industry paid $63 billion to the U.S. government, including federal, state, and local fees and taxes (Fig. 2.2).
- The use and availability of wireless telecom services and products created a $157 billion consumer surplus, which is the difference between what end users are willing to pay for a service and what they actually have to pay.

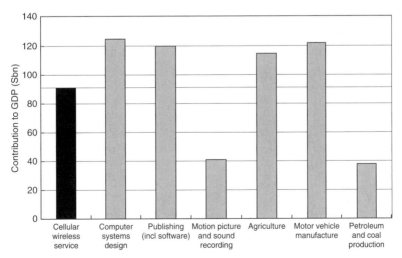

Figure 2.2. Cellular wireless services contribution to U.S. GDP (2004). *Source*: The impact of U.S. wireless telecom industry on the U.S. economy. A study for the CTIA, September 2005.

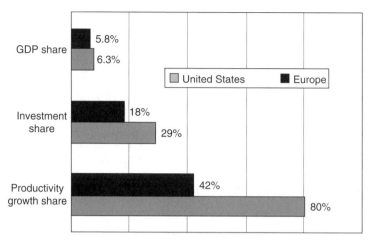

Figure 2.3. European Union vs. United States: ICT contribution to GDP, investment, and productivity growth (2004). *Source*: The impact of U.S. wireless telecom industry on the U.S. economy. A study for the CTIA, September 2005.

- On the basis of its conservative assessment of no additional services beyond what were available in 2004, Ovum predicted that over the next 10 years, the U.S. telecommunications industry would create an additional 2–3 million new jobs, add a cumulative $450 billion in GDP, create another $700 billion in consumer surplus, and provide cost savings of more than $600 billion to U.S. businesses.
- In 2004 wireless data services contributed only $8 billion in productivity benefits to the U.S. economy—roughly the size of Bahrain's entire economy.
- By 2015 these benefits have grown to more than $80 billion per year, which is approximately as big as the economy of Chile or the Philippines.
- All benefits to companies are actually savings from reduced time—not from increases in sales (Fig. 2.3).

Ovum also concluded that ICTs have had a much greater impact on labor productivity growth in the United States than in the EU. Therefore, while ICTs contributed 6% to GDP in the United States, they contributed 29% to investment and 80% to productivity growth (vs. 18% and 42% in the EU) (Figs. 2.4 and 2.5).

WIRELESS VERSUS WIRED NETWORK ECONOMICS

Wireless networks do not require the physical deployment of conduit and cabling to carry telecommunications traffic. This unique attribute will continue to gain advantage over the relatively static costs of construction and installation for wired facilities.

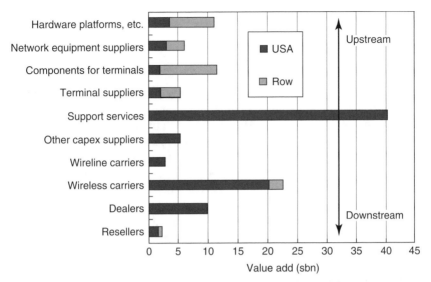

Figure 2.4. The GDP impact of the U.S. wireless industry. *Source*: The impact of U.S. wireless telecom industry on the U.S. economy. A study for the CTIA, September 2005.

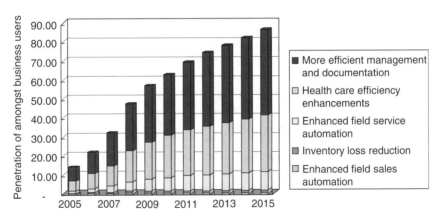

Figure 2.5. Productivity impact of wireless industry. *Source*: The impact of U.S. wireless telecom industry on the U.S. economy. A study for the CTIA, September 2005.

Advances in microprocessor power will continue to improve the capacity and efficiency of wireless networks disproportionately over the nonscalable manual labor–intensive elements of wired networks. We will have to bear with continually increasing costs of trenching, installing conduit, and pulling cable rather than continuing cost reductions of wireless equipment in parallel with improved efficiency and capacity. The end result will be a continuum of economic advantage toward the wireless architecture.

Cable system operators and installers of metropolitan and suburban fiber-optic networks to a node or premise typically describe their capital expense costs in terms of "homes" or "dwelling units" or "revenue locations" passed. The penetration of the deployed service in a given market defines the capital expense per subscriber. With the wireless system, the radio frequency "cloud" covers virtually all locations under the footprint of a given base station, thus dramatically reducing the cost per home passed to a level that allows for profitable operations even with modest market penetration—an advantage that wired networks simply cannot match in terms of capital efficiency for coverage. Offsetting the ease of coverage advantage held by wireless, wired coax or fiber systems will enjoy a long-term sustainable advantage in terms of their ability to support bandwidths far beyond the current capacities of wireless systems, with the unique exception of free space optics and the new systems exploiting the massive spectrum resources above 60 GHz.

BROADBAND MATTERS

We have established telecommunications' important contribution to the economy and society at large. Now let's take a deeper look at the evolving trends and how they impact on the adoption of technologies and in turn affect the information society. Over the past six to seven years, the impact of broadband on the information society has been significant. The availability of a transport that allows information and content to be accessed faster and cheaper has had a significant impact on how people use and distribute information across all types of applications and services. A recent study [4] done to measure the impact of broadband on the economy concluded that broadband access enhances economic growth and performance and that the assumed (and oft-touted) economic impacts of broadband are real and measurable. The communities in which mass-market broadband became available experienced more rapid growth in employment, in numbers of businesses overall, and in businesses in IT-intensive sectors.

As mentioned earlier, the availability of broadband empowers users to engage more efficiently with information, and more so than on slower networks. A study by Pew Internet Project [5], the research concluded that for broadband users, the always-on high-speed connection expands the scope of their online activities and the frequency with which they do them. It transforms their online experience. This led to a steady growth in broadband adoption among Net users. This pattern is quite evident in Fig. 2.6.

SO WHAT OF BROADBAND WIRELESS IN ALL THIS?

With the advent of EV-DO and WCDMA, the cellular technologies or the wireless WAN technologies that provide mobile wideband access (e.g., EV-DO speeds range

Figure 2.6. How broadband influences behavior. *Source*: The Broadband Difference: How online Americans' behavior changes with high-speed Internet connection at home, Pew Internet & American Life Project, 2002.

from 400 to 700 Kbps and with new enhancements (Release A and B) that will deliver multi-Mbps true broadband speeds that are higher than the most prevalent wireline broadband technologies such as DSL. In its landmark presentation *World Information Report* [6], the International Telecommunications Union (ITU) and United Nations (UN) identified two distinct paths to an information society (Fig. 2.7).

The role of mobile communications and the introduction of high-speed technologies such as 3G make wireless a key component of the information society. The ITU study came up with performance indicators to measure the Digital Opportunity Index (DOI), which is a tool for measuring progress toward building an information society.

Almost all the indicators chosen for the DOI have a mobile component. Mobile coverage and mobile subscribers explicitly relate to mobile communications, while others are embedded in indicators such as computers (e.g., smart phones, personal digital assistants [PDAs]) or Internet subscriptions (which can include mobile Internet subscriptions). The DOI can thus be split into fixed versus mobile technologies (Fig. 2.8). This allows analysis of each country's path toward

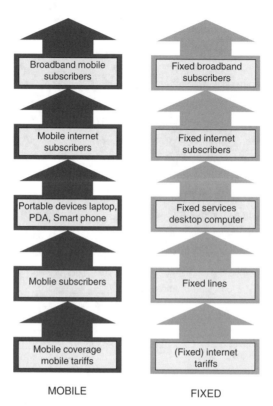

Figure 2.7. Two paths to the information society. *Source*: World Information Society Report, ITU, August 2006.

the information society. Evidence from country case studies and the trend toward ubiquity suggest that countries should not sacrifice one path at the expense of the other, but they should pursue both simultaneously. On the mobile path, broadband wireless plays a crucial role. In calculation of the DOI, mobile broadband penetration is an important contributor (Fig. 2.8).

HOW DOES THE UNITED STATES RANK AGAINST THE REST OF THE WORLD?

The United States has been behind other developed countries in broadband rollout and adoption. Michael J. Copps, current commissioner of FCC, lamented in a recent *Washington Post* op-ed piece [7],

America's record in expanding broadband communication is so poor that it should be viewed as an outrage by every consumer and businessperson in the

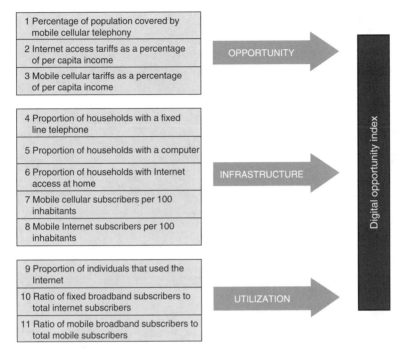

Figure 2.8. Digital Opportunity Index performance indicators. *Source*: World Information Society Report, ITU, August 2006.

country. Too few of us have broadband connections, and those who do pay too much for service that is too slow. It's hurting our economy, and things are only going to get worse if we don't do something about it. … The stakes for our economy could not be higher. Our broadband failure places a ceiling over the productivity of far too much of the country.

The adoption of 3G in the United States started in earnest in 2005 and has been growing rapidly. As of 2007, there were almost 64 million 3G subscribers [8] in the United States, primarily from EV-DO services of Verizon and Sprint Nextel. Korea and Japan, who had a head start with 3G deployment, have already reached over 80% penetration. In the United States, we are at a critical juncture of the inflection curve—the cusp (Fig. 2.9). All the growth conditions* were in place for the mass-market consumption to take off in 2007.

There is an emerging consensus about 3G among the analyst community. Lehman Brothers, in their report *3G Inflection* in 2006–2007 noted, "We believe

---

* For any wireless technology standard to succeed and get on a growth trajectory, there are three key elements that need to be in place—nationwide network coverage, cheaper selection of handsets, and healthy ecosystem of developers, application, and content providers.

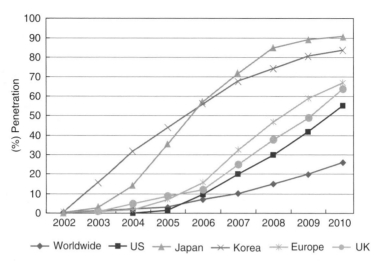

Figure 2.9. 3G subscriber growth forecasts. *Source*: Chetan Sharma Consulting, Global Wireless Update 2007, March 2008.

stable 3G networks, competitive handsets; a broadening portfolio of applications, and increased carrier marketing initiatives should drive a long awaited inflection in 3G demand in 2006 and 2007" [9]. The report also notes that "carriers in the US are migrating to 3G primarily for higher-speed data rates rather than lower-cost voice service" [10] Similarly, CIBC [11], Morgan Stanley [12], and others have indicated that their research shows a rapid increase in 3G adoption in 2006–2007. The fact that it took so long for us to get started shows up in the world rankings. The ITU and the UN released their updated landmark findings *World Information Report* [6], according to which the DOI for the top 25 economies in 2007 ranked the United States twentieth (Fig. 2.10). The ITU study came up with performance indicators to the DOI. If we plot the DOI against the broadband penetration ranking [13], we see the disparity between the United States and other leading economies (Fig. 2.11). The lower the ranking, the higher the DOI and broadband penetration.

So, why is broadband wireless important? As we saw in the Pew Internet Project results (Fig. 2.6), broadband wireless lifts the usage considerably. In a recent survey done by M:Metrics, a market research company [14], in every single category, broadband wireless, because of its better user experience and bandwidth, scored higher than lower-generation technology, and had a higher penetration of usage (Fig. 2.12). These trends are consistent what we saw in Korea and Japan over the past few years and what we are observing in Europe, which is on the same growth trajectory as the United States in terms of broadband wireless growth. There is absolutely no doubt that broadband wireless will have a direct impact on society and on the economy. We already see strong evidence of early adoption and direct productivity benefits for both consumers and enterprise workers. The public

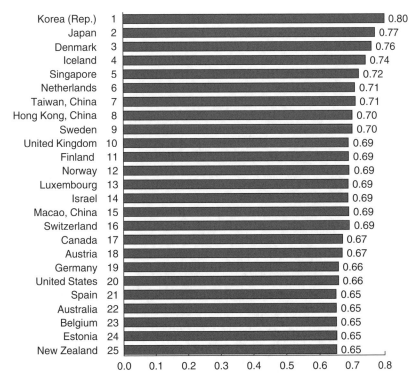

Figure 2.10. Digital Opportunity Index of Top 25 Economies, 2007. *Source*: World Information Society 2007 report, ITU, United Nations Conference on Trade and Development, http://www.itu.int/osg/spu/statistics/DOI/index.phtml.

at large is better informed because of improved user experience and broadband availability. As Michael J. Copps said in his *Washington Post* op-ed piece, "We need a broadband strategy for America. Other industrialized countries have developed national broadband strategies. ... The solution to our broadband crisis must ultimately involve public–private initiatives like those that built the railroad, highway and telephone systems. Combined with an overhaul of our universal service system to make sure it is focusing on the needs of broadband, this represents our best chance at recapturing our leadership position."

EXPANSION OF THE DOI

It is clear that telecommunications and wireless broadband as initially delivered over 3G wireless play an important role in a country's economy at the local, state, and national level. It helps foster a knowledge society, which has many other positive consequences. The World Summit on the Information Society initiated by

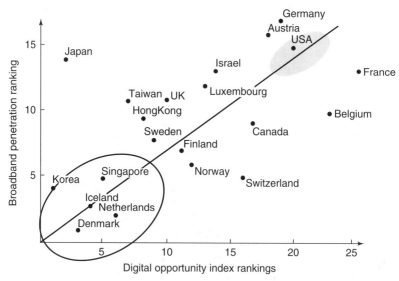

Figure 2.11. Correlation of Digital Opportunity Index and broadband penetration. *Source*: DOI—World Information Society 2007 report, Broadband Penetration Ranking—Organization for Economic Cooperation and Development (OECD), http://www.oecd.org/document/7/0,3343,en_2649_34223_38446855_1_1_1_1,00.html, Dec 2006.

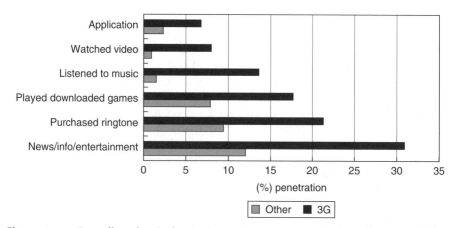

Figure 2.12. Broadband wireless's impact on user behavior. *Source*: Mobile Market Perspective, M:Metrics, June 2006.

the ITU and the UN made a strong commitment toward building a people-centered, inclusive, and development-oriented information society for all, where people can access and utilize information and knowledge. Furthermore, the *Geneva Declaration** contains a commitment to turn the digital divide into a digital opportunity for all and to provide access to ICT infrastructure and services that is universal, ubiquitous, equitable, and affordable [15]. To improve their competitiveness in the world, nations need to provide an accelerated plan for broadband availability. The availability of 3G and the expected growth in the next two years will go a long way in increasing the nations' DOI. Subsequently, we predict the same benefits will penetrate into the broader economy as we deploy true broadband wireless networks that extend all services, all the time, and at any location.

REFERENCES

1. Macklin B. The value of widespread broadband. Entrepreneur.com; Aug. 2002.
2. Entner R, Lewin D. The impact of the US wireless telecom industry on the US economy. A study for the CTIA—The Wireless Association; Sept. 2005.
3. TIA. Press Release; 2006 Feb 23. Available at http://www.tiaonline.org/business/media/press_releases/2006/PR06-17.cfm.
4. William Lehr, Carlos Osorio, Sharon E. Gillett (MIT); Marvin Sirbu (CMU). Measuring broadband's economic impact. A report for the U.S. Department of Commerce, Economic Development Administration; Feb. 2006.
5. Pew Internet & American Life Project. 2002. The Broadband Difference: How online Americans' behavior changes with high-speed Internet connections at home. Available at http://www.pewinternet.org/PPF/r/63/report_display.asp.
6. International Telecommunications Union. 2007. Digital Opportunity Index (DOI). Available at http://www.itu.int/osg/spu/statistics/DOI/index.phtml.
7. Michael J. Copps. America's Internet disconnect. Washington Post; 2006 Nov 8, p A27.
8. Chetan Sharma Consulting; March 2008.
9. Lehman Brothers. 3G Inflection in 2006–2007; 2005 Dec 20. Lehman Brothers, p 1.
10. Lehman Brothers. 3G Inflection in 2006–2007; 2005 Dec 20. Lehman Brothers, p 21.
11. CIBC. 3G Survey; Apr. 2006.
12. Morgan Stanley. The North American 3G Wireless Report; Feb. 28 2006.
13. ITU. Digital Opportunity Index (DOI). Available at http://www.itu.int/ITU-D/ict/doi/index.html.
14. M:Metrics. Mobile Market Perspective, M:Metrics; June 2006.
15. ITU. World Information Society Report; Aug. 2006.

* The first phase of the WSIS summit was held in Geneva in 2003, and hence the declaration title.

3

GLOBAL WIRELESS MARKET ANALYSIS

Wireless is the fastest-growing consumer technology of the last decade, and it is showing no signs of slowing down. In June 2007, the global mobile subscriptions eclipsed the 3 billion mark—over 50% of the population on the earth.* From expensive luxury devices of the early 1990s, mobile phones have evolved to be available for less than US$20 and include both voice and data capability. Along with the Web (aka the Internet), mobile communications has contributed to massive efficiency improvements in the world's economies. For example, for every 10% increase in subscriber penetration, there is a 0.6% impact on GDP. By aggregating GDP impact from national economies, the measured global impact could be over 8% [1]. From the early days of narrowband voice communications, mobile devices have morphed into devices with 1-GHz processor speed and 8 GB of onboard storage capacity,† enabling complex multimedia and communication applications that allow users to interact with information and entertainment

---

* This doesn't mean that half the population will have a mobile phone. This is for two main reasons: SIMs that are no longer used because of churn haven't been purged from the databases and multiple device ownership. In the most developed cellular markets, market penetration is well over 100%—Italy for example is over 140%. Clearly, this means that there are multiple connections per real user and some statistics are becoming available to measure this phenomenon. In Italy the number of SIMs per real user is around 1.7. The real number might be 1.2–1.3 meaning that actual subscriptions are 30–40% lower than that actually reported in these countries.

† Samsung launched the device in early 2007.

Wireless Broadband. By Vern Fotheringham and Chetan Sharma
Copyright © 2008 the Institute of Electrical and Electronics Engineering, Inc.

content. This chapter will take a look at the trends in various global markets to set the stage for discussions later on in the book.

MACRO TRENDS

This section reviews the trends that are common across the entire globe before delving a bit deeper into regional analysis. The last few years have been marked by exlosive subscriber growth in both mobile subscribers and data users. In terms of revenues, there is also a clear shift (in revenues generated) from voice to data services, even in low ARPU (average revenue per user) nations [2], with data becoming a necessary and consistently reliable revenue source that positively impacts the bottom line.

Massive Growth

The period 2003–2008 has been one of highest growth in the wireless industry, especially due to the significant momentum in China and India. In this period, our industry will have added over 2.2 billion subscribers or over two-thirds of the total subscribers by 2008 (mature markets have an average mobile penetration of 75–120%, whereas immature markets are at 20–50% penetration). In June 2007, the total number of subscriptions eclipsed the 3 billion mark; the next billion subscriptions are expected to come within two to three years. Mature markets in Western Europe, North America, Japan, and Korea have reached saturation,* while India, China, and Brazil are exhibiting tremendous uptake. As of March 2008, India and China were adding approximately 8 million new subscribers a month, and the markets are just getting started.

Moving from Voice to Data

As cellular migrated from analog to digital and with the advent of text messaging (TXT) in the late 1990s, consumers started getting exposed to something different than voice—data services. Starting from Scandinavia and then moving to Western Europe, Japan, Korea, and North America, texting gradually became a very popular application. In fact, in 2007, a good portion (40–50%) of the data revenues came from messaging.[†] However, this has started to change, as discussed later.

---

* It should be noted that the home-market saturation has forced major carriers to seek overseas strategy to find new subscribers. Some ventures like DoCoMo/AT&T Wireless Vodafone/Japan have failed, but major operators continue to invest in growing markets like India, Latin America, and other Asian countries.

[†] MMS or multimedia messaging is also part of this equation. However, percentage of revenues coming from messaging has been decreasing steadily over the last two to three years. By end of 2006, most of the major carriers like DoCoMo, KTF, Verizon, Cingular, Sprint, SK Telecom, and KDDI reported 60–70% of their data revenues coming from non-TXT applications.

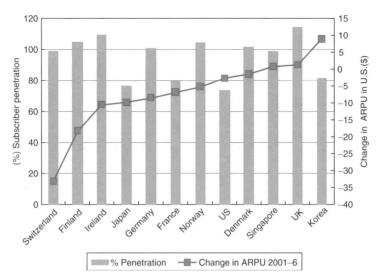

Figure 3.1. Percentage subscriber penetration and ARPU change in major wireless markets. *Source*: Chetan Sharma Consulting, 2008.

Pressure on Voice ARPU

It is a consistent trend in all major markets, from the developed markets such as Japan and Korea to developing markets such as India and China, that with increased subscriber penetration, there is pressure on voice ARPU. Voice has become commoditized over the past few years with rates dropping to US$0.01/min in India. As seen in Fig. 3.1, in most major markets ARPU has been under pressure between 2001 and 2007 despite a dramatic growth in revenue from data services. At the time of this writing, the growth in data services has not been able to stave off the free fall in voice revenue per subscriber.

Data as a Major Revenue Source

Carriers have been used to making billions every quarter from voice service revenues. It wasn't until the introduction of i-mode by NTT DoCoMo in Japan that mobile data services arrived on the international scene in earnest. The explosive growth of data services in Japan forced executives in carrier organizations to take the services seriously—at par with voice—and subsequently to make significant investments in the evolution of data services market. In 2007, NTT DoCoMo generated over $12 billion, 35% of its revenues, from its data services. In the United States, Sprint-Nextel, AT&T Wireless, and Verizon Wireless generate over $1 billion each quarter from data services. In China, where the ARPU is sub-$10 and data ARPU is $1–$2, the carriers generated over $12.2 billion in data

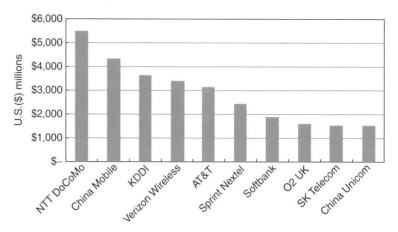

Figure 3.2. Top 10 global operators by mobile data revenues in 2007. *Source*: Chetan Sharma Consulting, Global Wireless Data Update 2007, March 2008.

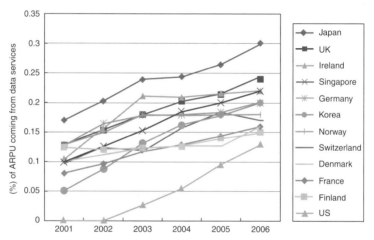

Figure 3.3. Data ARPU trends in developed wireless markets 2001–2007. *Source*: Chetan Sharma Consulting, Global Wireless Data Update 2007, March 2008.

revenues.* Other major carriers like KDDI, O2 UK, T-Mobile Germany, and United States reported in excess of $1 billion in data revenues for the year (Fig. 3.2).

As shown in Fig. 3.3, data ARPU is contributing in excess of double-digit percentage to overall revenue streams in all major industrialized nations.

---

*Japan and United States were ahead with $19.7 billion and $15.8 billion in data revenues for 2006, respectively. Top 10 operators generated a total of $46.8 billion from data revenues (*Source*: Chetan Sharma Consulting, more details at http://www.chetansharma.com/globaltrends.htm).

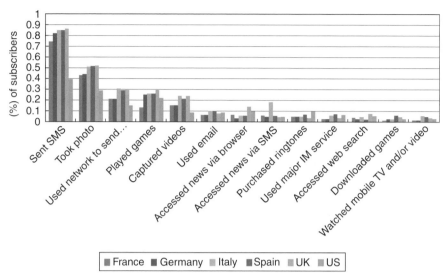

Figure 3.4. Mobile content consumption in major markets. *Source*: M:Metrics, February 2007.

What's Driving Data Growth?

Each market is different; the differences are dealt with later in this chapter. The key driver worldwide, however, is the introduction of a plethora of innovative and useful new data services, which is driving the growth in data usage. As shown in Fig. 3.4, text (TXT) usage dominates, while other forms of data services such as multimedia messaging (MMS), e-mail, games, browsing, search, mobile video, and mobile TV are starting to contribute to the data revenues as well. Japanese and Korean markets benefited from their relatively early introduction of 3G networks. These networks enable faster downloads and data transfer, thus stimulating more usage by the subscriber. More usage equates with more demand, which stimulates the creation and introduction of new services, a classic virtuous cycle that ultimately resulted in higher ARPU.

The main drivers for increase in content consumption are: the following.

Increase in Network Speed with 3G.* There is clear evidence that the introduction of 3G networks helps increase content consumption, primarily due to better network speeds, which results in better user experiences via quicker response times. Coinciding with these network improvements is the arrival of more powerful handsets, which enabled more enjoyable visual experiences, and the introduction of multimedia-rich applications. All of these improvements lead to more interactivity and an increase in the amount of time 3G subscribers spend

---

* EV-DO is evolution on the CDMA side with Rel 0, A, and B. On the GSM side, WCDMA, HSDPA, HDPA, and LTE form the evolution path.

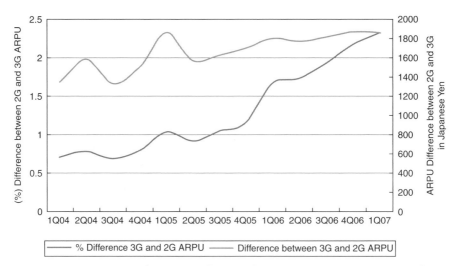

Figure 3.5. Data ARPU for 2G and 3G subs—NTT DoCoMo. *Source*: Chetan Sharma Consulting, 2008.

on their device; 3G has a direct correlation with increase in data usage and higher ARPU. In fact, 3G ARPU is generally two to three times the 2G ARPU (Fig. 3.5). NTT DoCoMo has the highest 3G subscriber penetration, and the difference between the revenues generated from the two network offerings has been widening since the introduction of 3G. Similar trends have been witnessed in Korea, Europe, and the United States.

By 2010, over 60% of the enterprise customers will be using 3G, both in North America and Western Europe. In Europe, UMTS/HSDPA/LTE (Universal Mobile Telecommunications Systems/High Speed Downlink Packet Access/Long Term Evolution) will be the primary standard, while in the United States, UMTS/ HSDPA and EV-DO Rev X will have almost equal market share. Both will provide theoretical downlink speeds of 100 Mbps and uplink speeds of 50 Mbps.

Device Capabilities: Faster Processors, Increased Screen Size, Different Form Factors.* From a functionality point of view, the capabilities of smartphones are getting better with every release. There are already 1-GHz processor mobile phones (from Samsung) and over 4-GB hard drives (from Samsung). Processing and storage won't be an issue by 2010. Display size will continue to be small; therefore, improvements in user experiences must be achieved to provide easy-to-use interfaces for all of the advanced services. Introduction of jog wheels, trackballs, and touch screens are a few examples of innovations that address the need for a better user experience. Things like voice recognition, which can be a

---

*The processor speeds are approaching 1 GHz, the screen resolution and size are becoming PDA-like on most new phones, and the industry has been experimenting with a variety of input modalities such as touch screen (iPhone), voice, etc.

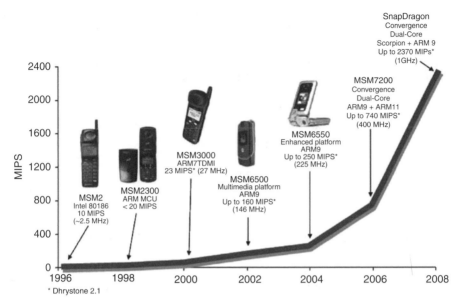

Figure 3.6. Qualcomm's MSM chipset roadmap. *Source*: Qualcomm.

decent experience with voice processing on a 200 + MHz device, complemented by analytics will help with better task orientation and completion. However, battery life is one area that hasn't kept pace with technology growth. Several carriers in Asia are looking at fuel cells, and it is quite likely that such batteries will start showing up in high end smartphones by 2008–2010. Figure 3.6 shows the evolution of processing power in Qualcomm's MSM chipset, which forms the basis of CDMA and many WCDMA handsets. The newest in the series is SnapDragon.

Key features of SnapDragon are as follows:

- Always On
- Low power consumption through custom CPU and DSP cores
- All the power of a laptop performance
- Superscalar CPU: Scorpion surpass 2100 DMIPS at 1-GHz speed
- Next-generation DSP running at 600 MHz
- High resolution VGA to XGA support for uncompromised video and computing ubiquitous connectivity
- CDMA, WCDMA, HSPA, GPS, Bluetooth, WiFi, Broadcast (MediaFlo, DVB-H, etc.)

Mobile devices include a primary CPU and coprocessing hardware to support multimedia applications. Hardware accelerators have the advantage of consuming low power and demanding low memory (both RAM and ROM). Their architecture is specifically designed to maximize the performance of a given application (e.g., audio, video, image, or graphic processing). However, these chips are less flexible in their use, as their design is based on hardware. Consequently, they are not upgradeable and do not allow applications to be reconfigured. Hardware accelerators are a good solution for cost-sensitive feature phones aiming to a specific segment of the handset market, such as onboard game phones, camera phones, or music-enabled phones. A common coprocessor in this market is the camera application processor (CAP).

Chip manufactures are improving their architecture, including power management system. ARM,* for example, has recently introduced ARM Intelligent Energy Manager (IEM) technology, which optimally balances processor workload and energy consumption, while maximizing system responsiveness for better end-user experience. In future-generation processors, manufacturers will focus more on reducing energy consumption rather than racing for higher speed.

Data Plans Getting Less Confusing and More Cost Effective for Consumers.[†] Though some consumers might argue otherwise, data plans are becoming less complex. Starting with Willcom in Japan, and then European carriers like 3 and Vodafone, flat-rate pricing irrespective of the applications being used is becoming the norm. The U.S. service providers are likely to resist the change in the short term,[§] but we are inevitably entering an era of flat-rate pricing with some tiered pricing for either quality of service (QoS) or premium/newly introduced applications and enhanced services. This evolution of the business model for data services helps the consumer, who no longer will have to watch the meter to see where he or she stands to avoid overage charges.

Software Drives Hardware. The introduction of iPhone at MacWorld in February 2007 intensified the role of software in a mobile device.[¶] For the past few years, smartphones have been slowly whetting our appetite for downloading and installing applications beyond the basic features. As the market penetration of such devices increases, more applications in the devices that interact with Internet content and services will be seen.

Device Replacement Cycles Are Shrinking. The device replacement cycles are constantly shrinking. In Japan and Korea, the mobile phone gets replaced in

---

*ARM Holdings plc is a leading provider of semiconductor intellectual property (IP) to the semiconductor industry. ARM is publicly listed on the London Stock Exchange (ARM) and NASDAQ (ARMHY).

[†] Both a la carte and all-you-can eat plans are available for consumers to pick and choose the services they like, though there is a sense of sensory overload.

[§] Helio—an MVNO started by Earthlink of the United States and SK Telecom of Korea introduced a flat-rate pricing. In mid-2008, Helio was acquired by Virgin Mobile USA. Even AT&T Wireless has some plans that allow you to a qualified unlimited data access and services such as messaging, video streaming, browsing are all bundled in.

[¶] Again, Japan and Korea were ahead in realizing the role of software in mobile devices and their handsets already took advantage of a number of features that Apple integrated into the iPhone.

less than 12 months, in the United States and Canada in around 17 months, and in Western Europe in approximately 16 months [3]. In fact, 2006 marked the first time in the history of the industry that replacement devices took over the new device sales, and by 2010, people replacing their mobile phones will account for 80% of the mobile phone sales. With this constant device change cycle, newer and better devices are getting into the hands of consumers (e.g., the 3G penetration reached over 80% by early 2008 at NTT DoCoMo). Driven by word of mouth and advertising, users want to try out the new content offerings in spite of the challenging navigation.

Mobile Media Offerings Are Increasing. Over the past two to three years, content has been evolving from ringtones and graphics and photo sharing to more advanced multimedia and interactive services such as streaming multimedia interactive games and applications. Mobile TV, and more particularly mobile broadcast video, is one of the most talked about new services for wireless subscribers for sometime.

South Korean SK Telcom's MelOn mobile music service was one of the first services to embrace the concept of ubiquitous music use, offering over 800,000 tracks anytime and anywhere for consumers to enjoy on the PC, mobile phone,or MP3 player. For a monthly flat rate of around US$5, SK Telecom customers can stream and download music online or via mobile, as well as reproduce their songs without limitations, as long as the monthly subscription is maintained. All tracks are protected by digital rights management (DRM). SK Telecom has over 20 million subscribers and music has become its biggest grossing data service. As of early June 2006, MelOn had nearly 4.5 million users, with 15% of those on a flat-rate subscription. According to SK Telecom, the MelOn service ranked first among South Korean paid music sites in terms of revenue.

In Asia, 2006 was a banner year for mobile music. Supported by the ongoing roll out of 3G networks and rich media content such as full audio and video tracks delivered wirelessly to handsets, millions of consumers experienced mobile music services. Asia accounts for 25% of global mobile digital sales, led by Japan, South Korea, China, and Indonesia. Mobile music accounts for an overwhelming 90% of Japan's digital sales and growth in the sector remained strong in 2006, pointing to a mature but not yet saturated market. Japan has an advanced "mobile culture" with high penetration of advanced phones and 3G.

Eric Nicoli, CEO EMI Group, said, "In this Internet age, the consumer is using music content more than ever before—whether that's play listing, podcasting, personalizing, sharing, downloading or just simply enjoying it. The digital revolution has caused a complete change to the culture, operations and attitude of music companies everywhere" [4].

Similarly, mobile broadcast TV takes phone to a different level. Again, Korea and Japan introduced these services two to three years ahead of rest of the world. Verizon launched Qualcomm's MediaFLO in 1Q 2007 offering eight live channels of exceptional quality. While the penetration of devices with such capability is still small, the potential of growth is anticipated to be substantial.

Mobile Operator Portals Are Opening. The level of openness varies by geography, with the United States being one of the most closed markets. Among the major carriers in the United States, Verizon has resisted the change the most. The argument that is often put forth is the difficulty in managing the network if things were open. While some of the concerns are well founded, initiatives like disabling Bluetooth and WiFi are actually counterproductive and draw the ire of users. In spite of the carriers, off-deck content is growing at a steady pace, and with mobile search being integrated into the carrier offerings and percentage penetration of smartphones on the rise, the vicelike grip of the carriers is slowly but surely slipping away. Things are quite different in Europe and Asia. In Europe, 60% to 70% of the content sold is actually off-portal. A major content provider like Disney is actually selling over 90% of the content off-portal in Europe. In Asia, carriers exert less control over access and provide more generous revenue share for the ecosystem with some exceptions.*

Context-Aware Applications Increasing. Mobile content started pretty much the way Internet content started—categories and hierarchical menus. However, things have been changing with the introduction of mobile search and location-based services (LBS). Mobile search is helping collapse the overly complicated menu structure. It has taken years to get here, and there is still much room for improvement. Location-based services provide context of space and time, thus making the interaction more relevant, providing a better user experience for the user and ideally making purchase decisions more efficient. Global positioning system (GPS) capability is starting to appear in the market, especially in commercial applications. In the United States, Sprint was the first one to open up its LBS APIs (application programming interfaces) to developers, albeit on a selective basis. Others are working to become more developer friendly. AT&T and T-Mobile that rely on network triangulation for LBS are further behind in deploying commercial LBS applications and services, although as GPS-enabled GSM handsets get introduced in the market, the GSM carriers will catch up.

The introduction of voice-based navigation by firms such as Voicebox and Nuance are futher helping in solving the usability problem with mobile.

Increased Involvement of Media and Brands in Mobile. The producers of the hit movie "300" (released by Warner Brothers) prepared promotional content for mobile community sites as part of their promotion strategy. Similarly, big labels have embraced mobile as a key component of their digital strategy. With the explosion of MP3-enabled phones, the impact on iPod and the like is palpable. In fact, the birth of iPhone is a recognition that mobile devices are the future of digital music and affirms the digital music consumption trends in Asia. EMI—one of the pioneers in digital music initiatives—is even bold enough to become the first

---

*To follow the detailed analysis on the subject, read Open Gardens blog by Ajit Jaokar.

label to take up Steve Jobs' challenge of DRM-free music content.* EMI expects that digital revenue could account for up to 25% of total recorded music revenues by 2010, and mobile will form a good percentage of that mix. All major brands are looking to either extend their content catalog to this new medium or touch the customer via the mobile channel.

Convergence. Double, triple, and quad plays are all about bundling of services to increase the lifetime value of the customer and reduce churn. As the competitive pressures are increasing, cable/MSOs, Telcos, and wireless operators are planning on offering triple and quad plays either through direct investment (like Verizon investing in FTTx) or partnership (like Sprint's partnership with cable companies) or acquisitions (like NTL acquiring Virgin Mobile in the United Kingdom). Companies like AT&T who have all the network pieces in place are actively pursuing their "three-screen" strategy.

The benefits of the bundle are lower churn, increased ARPU, and higher customer satisfaction ratings. The combination of these benefits creates a valuable customer, one who is likely to be more loyal than the average customer.

Different content and applications markets are at varying levels of maturity with respect to factors discussed above, with Japan and Korea ahead in almost all categories followed closely by Europe and North America and then rest of the world (including India and China†). There is clearly an increasing awareness that the overall usage and consumption of content is showing tremendous growth all around. There are, however, significant regional differences reflecting cultural preferences within specific markets. Thus, as we enter an era of mass specialization, there will still be plenty of room for variety and unique attributes reflecting the special needs and desires of users within their local social and market fabric.

THE ERA OF "MASS SPECIALIZATION"

Future Predictions and Areas for Creative Business Development

The "Long Tail" Market. Chris Anderson popularized this description of the "mass specialization" market concept of the coming era. As traditional mass markets are fractured by the disintegration of mass media and popular culture yields to special interests that can be served to meet every individual's personal

---

* "The Good, The Bad & The Queen," became the first DRM-free EMI album launched on 4th April, 2007 allowing fans the option of choosing to play music across a range of devices and platforms, including digital music players, mobile phones, and home music systems.

† Even though phone prices in India and China are among the lowest anywhere in the world with sub-$20 handsets, they are still feature rich, capable of browsing, messaging, and even video streaming. In fact, Reliance Infocomm, the largest carrier at the time, launched its CDMA line of handsets with a multimedia player on every single handset, something that took their Western counterparts years. Though the performance of video was limited to 1–2 fps at the time (2002), it demonstrated the direction Reliance wanted to go. Some of the industry's firsts, like streaming of a full-length feature film (Bollywood), actually came from India (Bharti launch).

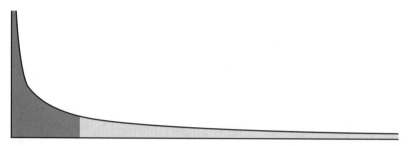

Figure 3.7. The long tail market (represented by the lighter shaded area).

interests. In a world where the Internet eclipses broadcast radio and television and newspapers as the information source of choice for its users, the traditional concept of mass-market power will be replaced by the mind share and purchasing power of a large number of individuals participating in a vast array of "micromarkets," which in the aggregate eclipse the volume of the traditional "mass market." (Fig. 3.7)

The potential for long tail products emerging in the wireless broadband industry is highly likely. As the global penetration for these presently mass-market products continue to expand and commoditize over the coming years because wireless penetration extends beyond its current reach of 3.5 billion (as of late 2008) users worldwide, the demand for product differentiation and niche market products will grow. If only for the potential to maintain margins on handsets, the migration to "mass specialization" is inevitable. How rapidly the commoditized mass market for cellular handsets extends its "long tail" into the future will depend on the speed with which the manufacturing industry can implement short-run, software-defined product development and manufacturing techniques that will allow customers to customize their product purchases from the equipment vendors.

Personalization of Hardware

The authors anticipate a future market for the design, manufacture, and delivery of personalized communications devices. Consider the potential of a Web-based environment for the purchase of devices, which are manufactured using flexible assembly lines to deliver "mass customization" of consumer electronic devices. Envision an environment where "Dell competes with Samsung via Flextronics with cost offsets supplied by Google in exchange for registering for personalized advertising push preferences...."

Initially mass specialization manufacturing will provide customer control over the incorporation of features, functions, storage capacity, processor power, battery choice, form factors, color, screen size and resolution, accessories, and software loads. Ultimately even the form factors of the enclosures will be open for variety, as computer-aided design (CAD) and computer-aided manufacturing (CAM) are joined by the growing industry of advanced low volume plastic-manufacturing techniques.

These short-run manufacturing technologies currently are being pioneered with 2Gs and 3Gs of the following prototyping technologies presently in use.

- Stereo lithography (SLA)
- Fused deposition modeling (FDM)
- Selective laser sintering (SLS)

Mass Customization

When referring to mass customization the authors are referring to the ability of individual customers to modify the appearance and the "look, sound, and feel" of the UI. We are exiting an era when the mass customization potential for cell phones was largely limited to ringtones and display "wall paper" and entering into a period when virtually all aspects of our wireless handsets could be customized by the owner of the device. As discussed above regarding the personalization benefits from the integration of flash memory into cellular handsets, personalization has become a significant subindustry serving the cellular market. The increasing power and performance of handsets enables users to leverage simple software tools to create new display graphics, or download ringtones and other multimedia enhancements that are highly personal and unique to them.

The ability to select special colors, texture, and other surface treatments for cellular handset cases, or "wrappers" as they are sometimes referred to, will also take on added importance as users continue to seek ways to individualize their possessions.

In addition, because of the incorporation of low cost, high capacity storage into small form factor electronic devices, service providers and content developers are able to provide either prestored content that is accessible to users using an access key purchased from the rights holder or, alternatively, real-time (or time-shifted) downloadable applications either from the network service providers' own inventory ("walled-garden" content) or in an open system directly from the public Internet.

The addition of lanyards, key fobs, antenna options, and "pulse" lights that flash during conversations are extremely popular cell phone accessories in Asia at the present time. At the upper end of the phone fashion phenomena are the diamond-encrusted and gold-plated handsets that crossbreed consumer electronics with high fashion and fine jewelry. Each of these examples demonstrates the desire for human beings to personalize their possessions and to enjoy participating in the never-ending game of fashion one-upmanship. As true broadband is added as a feature to the phones of the near future, the novelty of pure speed will wear off quickly and the bandwidth will be applied by creative practitioners to deliver new forms of customization to the wireless communications business.

Outsourced Electronic Manufacturing Services

The assembly of mass-market consumer electronic devices, including most of our wireless infrastructure and handsets, largely takes place in Asia. This trend started in the 1960s in pursuit of cheap hand-assembly labor. Over time, the need to source components near to the point of assembly resulted in a high value-added sector of the electronics industry also growing deep roots in the Asian region. As we have moved forward into an era when the amount of hand labor has reduced to less than 1.5% of the cost of an electronic device and our products are largely assembled using robotic equipment, we are essentially stuck with Asia as the default manufacturing region for our electronics, as most of the component suppliers have established local presence in the region that would be nearly impossible to replace.

The outsourced electronic manufacturing services (EMS) industry has grown and matured through consolidation among many of the leading EMS companies in the space, resulting in the dominant presence of Flextronics. It is the leading provider of advanced product design and EMS to many of the leading original equipment manufacturers (OEMs) that address numerous mass markets for their products, including the computing industry; mobile communications handsets; consumer electronic products; telecommunications infrastructure; commercial and industrial products, including white goods, automotive, marine and aerospace, and medical devices and components.

In addition to its EMS business lines, Flextronics has been providing value-added services in supply chain management and warranty administration and has recently begun moving into the market for complete product development services. This sector is called the original design manufacturer (ODM) business. Through acquisition of industry-leading industrial design and product development firms, Flextronics is positioning to further increase the value of its services to the major OEMs throughout the world. It is one of the world's largest EMS providers, with revenues from continuing operations in fiscal year 2007 of $18.9 billion. It has developed a vertically integrated campus approach to deliver its services to many of the world's largest OEMs. As a true global supplier, it participates in the value and supply chain throughout the life cycles of the products it makes. As the market continues to evolve for the development of flexible, short-run, specialized product manufacturing to create and deliver products that address the aggregated demand among the emerging "long tail" market for the products, we will see the emergence of "mass specialization."

As the number of OEMs that outsource their manufacturing services to EMS companies over the past decade increases, many opportunities have emerged for them to grow rapidly through mergers and acquisitions or the former manufacturing assembly facilities of organizations that are now served by the EMS outsource suppliers. Continuous downward price pressure on the manufactured products made by the EMS vendors has resulted in the concentration of their plants in low cost regions, including China, India, Malaysia, Mexico, and Eastern Europe.

Flextronics pioneered the unified, vertically integrated campus approach to the OEM electronics manufacturing industry. As the outsource manufacturer for many of the world's leading brands, Flextronics has developed the purchasing power and the in-house tools for quality and manufacturing control that would be almost impossible for smaller organizations to match in terms of cost and efficiency and quality. The next step for this key player in the electronics equipment value chain is to move from an OEM role into a more proactive ODM role. Adding the design and development engineering talent to its existing expertise in manufacturing efficiencies will enhance the overall revenue of the company and increase its participation in far more of the overall value chain.

Personal Customization of Services

As we move into an era of open network systems, some of the promising, but too early, Internet concepts will take hold again to extract value from the wireless broadband ecosystem.

The application service provider (ASP) business model that was gathering much momentum before the Internet bubble burst in year 2000 will be augmented by its more recent progeny, software as (a) service (SAS). In the SAS model, whole business processes can be outsourced to third parties, hosted off-site with no distance sensitivity, leveraging software interfaces and applications to automate the process. The arrival of the true broadband wired and wireless Internet is legitimizing many of the earlier failed visions that were anticipated to revolutionize computing and network communications. The so-called network computer proposal by Oracle, which was essentially a diskless PC, was rendered essentially pointless under the context of the immature broadband network, as the limitations of the access bandwidth of those (circa 1996) times simply rendered the whole concept impractical. As the ubiquitous wireless broadband Internet comes online, many of these "old ideas" will be resurrected and implemented within the framework of the new low cost broadband service offerings. How and by which players the disruptive aspects of all the new low cost access capacity will be tolerated, resisted, or assimilated will identify the most likely winners and losers among the service providers facing these challenges to their historical way of doing business.

Subscribers are going to be seeking services that match directly to their personal requirements and have limited time or tolerance for nonresponsive service providers. Failure to embrace open and adaptable service delivery platforms will be the death knell of legacy players who simply cannot adapt fast enough to the realities of the new generations.

User-Controlled Real-Time Carrier or Service Provider Churn. As users gain control over their platforms, their ability to load and run the specific applications that they want to run and to reach any content hosts that are on the Web will lead to another new by-product of the mobile user empowerment movement. The authors anticipate that users will develop preferences among a plethora of content and applications providers and that they will revolutionize the concept of churn to reflect real-time choices made by users to switch from one service provider to another in search of their favorite source for a given type of content (e.g., VoIP, movies, or music). The industry will also be exposed to the power of the "auction market" wherein the authors anticipate that there will be attempts to sell fallow capacity to the "highest bidder" in real time, using the maturing Internet tools that have been refined and polished by eBay and others.

Edge Caching (Mass Storage Meets Fallow Network Capacity)

The continued progress in the area of materials science and data storage products that leverage these technologies into ever more efficient and high density solutions has resulted in remarkable increases in capacity and reductions in device size.

The proliferation of the iPods feature embedded data storage devices of typically 30 or 60 gigabits of storage capacity on disk drives no larger than U.S. 25 cent coin. As the storage capacity in network edge devices increases, the applications that can benefit from having massive low cost storage available in the hands of users will proliferate. We anticipate that the network operators would be able to download massive amounts of common information such as news feeds (both real-time and daily reports), magazines, books, videos of all types, and software applications or content into participating user devices. These downloads will occur during both the off-peak hours of the network utilization and within unused gaps in the data transmission loads that appear during the peak-hour operations. In addition, the ability for device manufacturers to load a large amount of embedded optional software and content onto new user devices is also a by-product of low cost, highly distributed mass storage devices.

Targeted Advertising

The reciprocal side of user-defined customization for features and personalization of its wireless broadband services is the growing ability of the advertising industry to deliver highly targeted advertising messages, tailored to individual consumer preferences. As the success of Google, Yahoo, AOL, and other major participants in the online advertising industry has proven, the value of advertising that is targeted to the specific interests of individual consumers can garner significantly higher rates per impression compared with the "shotgun" advertising techniques that defined the mass-market environment of the past.

How the advertisers identify appropriate candidates for specific advertising messages or information is a fertile field for development. Previously, the use of "privacy policies" that were often better described as "privacy invasion policies" established by Internet portal operators and advertisers were used to acquire the filtering required to identify the purchasing and "click-to" interests of specific consumers. A growing awareness of the abuses have occurred in the domain of "spyware" and spam generators, as our industry moves from a "brute force" approach to targeted advertising to more nuanced and less intrusive techniques to connect the advertisers with appropriate consumers who have demonstrated a proclivity to be interested in whatever is being promoted. The techniques that are being commercialized at present have been designed to protect the anonymity and privacy of the specific individual's name and identity. One of the leading pioneers in this next wave of targeted advertising is Feeva Technology, Inc. (Feeva), which has been developing a new technique that links specific user behavior with advertisers and content owners who have messages or offers that should be of interest to the specific user. At the time of this writing, Feeva is still operating in a "stealth mode" and the authors refer readers to its Web site to obtain an update of its progress in matching consumers with advertisers of prefiltered interest while retaining the privacy and anonymity of the consumer.

Push and Pull Customized Advertising Models

The delivery of targeted advertising to specific individuals follows two parallel trajectories. The first is probably the highest value-added approach for advertisers, and it is the "opt-in" approach where specific consumers register their interest in receiving information about products or services that they have identified through either an online or an off-line (mail or physical questionnaires) process by filling out "bingo cards" or other physical registration forms. The opt-in approach has historically been leveraged by both e-mail and direct-mail advertising campaigns and solicitations. It has also been tied to specific interaction with a given Web site or portal with which the consumers have registered their preferences.

The second form of targeted advertising technology does not require the opt-in of specific consumers, and which gathers and stores the preferences and buying histories of specific consumers through various data mining exercises, depending on the sophistication of the advertiser or advertising host such as Google. A growing body of privacy and consumer protection statutes thankfully intrude on the unfettered gathering and application of the targeted consumer data that is valuable to advertisers in focusing their resources on only the most appropriate candidates for reception of their advertising messages. However, it is within the context of sampling and archiving specific consumer behaviors that will hold the most value for advertisers in the future of the ubiquitous broadband marketplace. How advertising messages are disseminated to combinations of both legacy mass-market audiences and the emerging long tail markets, which require highly targeted messaging, will be one of the largest business opportunities that will emerge out of the coming bandwidth and open network environments over the coming decade.

Personal Real-Time Context (as Contrasted to Just LBS)

One of the ultimate tools for the future of "mass specialization" applications will be the integration of "context awareness" into the personal broadband wireless devices of network users. The contributors to the creation of context-aware capabilities as an inherent feature of next generation of broadband wireless networks are both passive and active. Location sensing is the foundation of the emerging "context engines" that will extract and integrate the location-based information of specific users and their devices at a given location. However, context requires far more information than just location. When position coordinates are cross-correlated with stored database information regarding points of interest that are present at or near any given location, the next level of context awareness is attained. The ability to be alerted to localized information or services that the user has identified as of interest and the filters that sift through extraneous data that are not of interest at the instant location is key to delivering the utility value of context LBS.

The next layer of context information requires instruction by the individual user to establish the preferences and filters required to winnow out contextual information that is of little or no interest to the user. These resulting "rules tables" will be applied to the context engine to refine the information search that is customized for each individual consumer's preferences. As our user devices continue to increase in processing power and our software developers progress with the application of artificial intelligence (AI) to assist in the accomplishment of tasks that require real-time decisions based on information less defined than the traditional "rules tables," the context engines will begin to reflect a tremendous amount of learned behavior based on the prior activities of the user. Thus, the knowledge of what and how to function within the context of being "at work" will alter substantially when the context engine is aware that its user device is "off duty" and that the contextual expectations and tasks will be modified to reflect the change of context. Increasing dynamism and the ever-expanding intelligence of our computing devices will ensure that the ultimate development of "mass specialization" will be the incorporation of contextually aware, completely generic communications and computing devices, that are transformed dynamically, in real time, into a wide range of specialized applications or missions.

Appear Networks of Stockholm, Sweden, has been one of the early pioneers in the context-aware market for improving the efficiency of individuals and the organizations within which it works. It has been concentrating on the application of contextual awareness primarily to enterprise operations. Early success of Nederlandse Spoorwegen (NS), the National Dutch Rail Authority, was acknowledged by its winning the "Best Mobility and Wireless Project of the Year" award at Cisco Networkers, Europe 2007. The Dutch Rail project was the first deployment of a wide-area implementation of a WiFi wireless broadband network that linked approximately 10,000 handheld computing devices, which were issued to frontline employees throughout the organization to support all of the administrative and dispatching requirements to match resources with present location. Appear Networks has continued to find numerous applications for its context-aware LBS middleware technology platform with a growing range of large-scale enterprises.

REVIEW OF MAJOR MARKETS

As you might have gathered by now, or known from your own experience in the industry, different markets are progressing at varying levels of maturity. Figure 3.8 gives the level of data revenues in 35 major markets as of 2006, and these numbers are increasing every quarter. This section briefly looks at the state of data services in some of the major global markets (Fig. 3.9).

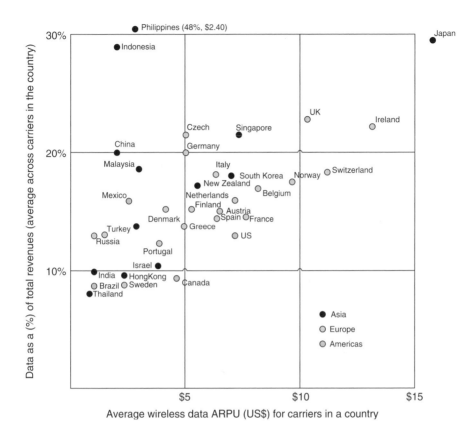

Figure 3.8. Contribution of wireless data revenue to the overall ARPU for major nations. *Source*: Chetan Sharma Consulting, Global Wireless Data Update 2007, March 2008.

Korea and Japan

Japan and Korea remain the world's two most advanced mobile markets. For the past decade, both markets have been at the forefront of innovative mobile data applications and services. Both deployed 3G in 2001–2002, and in early 2008 their 3G penetration was above 80% and is expected to reach almost 100% by 2010. Better networks and devices have helped keep the ARPU high, although carriers in these markets have started seeing some pressures of commoditization and saturation.

China and India

While Korea and Japan are the most advanced mobile markets, China and India hold the most potential of going forward. The growth in both the Chinese and Indian markets has been stunning, although it should not come as a surprise given that every third human being on the planet comes from these two nations.

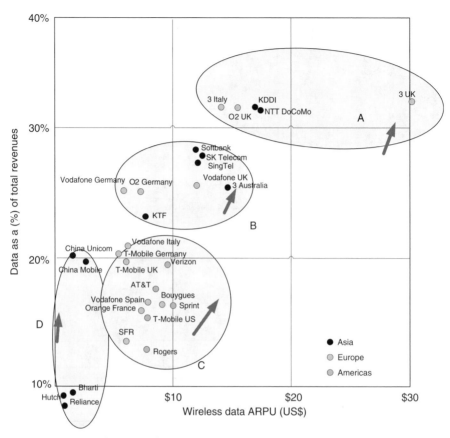

Figure 3.9. Contribution of wireless data revenue to overall ARPU for major carriers. *Source*: Chetan Sharma Consulting, Global Wireless Data Update 2007, March 2008.

The market for mobile phone service in developed countries is more or less saturated. There are only two major paths for further mobile phone industry growth: the industry can pursue new subscribers in developing countries, such as India and China, or sell value-added services to existing subscribers primarily in developed countries.

China has over 560 million mobile subscribers (as of Q1 2008), which is more subscribers than the combined population of United States and Canada. China's number one carrier, China Mobile, is the world's largest and most valued carrier. The average ARPU is around $10 with data contributing approximately 22% of this revenue figure. For the past several years, China has been struggling to decide on its 3G strategy. While the domestic trade industry is pushing for its homegrown standard of time division synchronous code division multiple access (TD-SCDMA), no one outside of China thinks much about it, even though it has been incorporated under the 3GPP standards umbrella. China thinks its huge domestic market should be sufficient to ensure that its domestic standard reaches full-market deployment. The development of TD-SCDMA from its fixed wireless genesis into a fully mobile solution has been plagued by numerous technical challenges and regulatory uncertainty.

China, in addition to being the largest market, is also the biggest promoter of Linux-based mobile devices. Adoption of Linux as a major handset operating system (OS) will tilt the dynamics of the top three smartphone OSs (Symbian and Windows Mobile being the other two) in the next few years.

In 2006, India passed the 100 million—subscriber mark, and in doing so, it surpassed Japan and Russia to become number three in terms of the total subscribers. It has been experiencing such incredible growth that, by late 2007, it had surpassed China in average net-adds per month yielding 8–9 million new subscribers per month versus China's 7–8 million subscriber adds per month. In March 2008, India crossed United States as the number two market in the world. Large carriers such as Vodafone returned to the market and all major manufacturers have set up both R&D and manufacturing plants in India to take advantage of this boom (Fig. 3.10).

India offers important lessons on how the mobile phone industry can increase teledensity in developing countries. More than 75% of India's roughly 1.1 billion people still do not have telephones (wireline or wireless).* However, the number of wireless subscribers added each month in India has climbed rapidly and is now at 8 million (as of mid-2008).

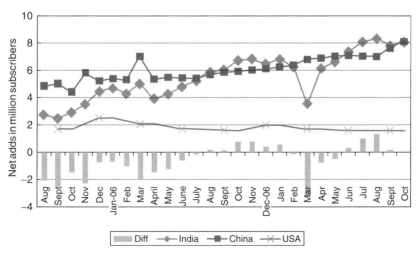

Figure 3.10. Monthly net-adds: India Versus China. *Source*: Chetan Sharma Consulting, Global Wireless Data Update 1H 2007, September 2007.

* As of 2007.

The Indian wireless market ARPU is around \$7–\$8. With airtime going for just 2–3 cents/min, however, the average Indian subscriber can talk for 500 min/month—comparable with what mobile phone subscribers in the United States receive for four to five times the cost. India's mobile phone subscribers cannot afford to pay as much as the U.S. consumers, but they expect comparable service. In addition, while India requires low cost handsets, the low end market is segmented; India's consumers demand a choice of handsets at different price levels.

Four factors are driving the growth of mobile phone services in developing countries: (1) the declining cost of mobile phones; (2) the availability of more efficient and less-expensive mobile phone network infrastructure; (3) the continuous entry of broad swathes of the populace into the middle class; and (4) the increasing business, social, and entertainment utility of mobile phones.

India is a country of villages, and penetrating the rural market is essential to raise teledensity to developed country levels. The challenge is to deliver cost-effective mobile phone service over a wide geography.

Europe

In the 1990s, the development and proliferation of GSM emerged as a model for cooperation between standard-setting bodies, governments, and industry. The deployment of GSM brought together European nations and vendors from around the world—a process that deemed an unequivocal success worth repeating. While the United States struggled with multiple standards and Japan isolated itself with proprietary technologies, Europe deservedly reaped the benefits of a common standard.

The ITU* sought to replicate the success of GSM as a 2G technology by developing a common standard for 3G mobile communications, heading off the challenges of global roaming and interoperability; it failed miserably. The absence of a common 3G standard[†] and the combination of poorly designed spectrum auctions and the bursting of the telecom and Internet bubbles left Europe's wireless industry in a precarious situation. Intoxicated with the success of GSM, some European regulators pressured carriers into paying exorbitant sums for 3G spectrum[§] and locked themselves into the WCDMA technology platform. In the midst of this turmoil, non-European nations continued to charge forth with upgrades to 3G infrastructures and next-generation technologies such as Mobile

---

*The ITU is unique among international organizations in that it was founded on the principle of cooperation between government and the private sector. With a membership encompassing telecommunication policy-makers and regulators, network operators, equipment manufacturers, hardware and software developers, regional standards organizations, and financing institutions, the ITU's activities, policies, and strategic direction are determined by the industry it serves.

[†] WCDMA and CDMA2000.

[§] European carriers paid in excess of \$100 billion for 3G spectrum. In the United Kingdom, five companies committed \$35.4 billion in winning bids. In Germany, six companies committed \$45.85 billion.

WiMAX/WiBro emerged in the United States and Asia, while Europe's wireless technology and market lead evaporated.

The broadband wireless market started to return to life around 2005, with most carriers finally deploying 3G networks in a meaningful way, and subscriber data usage started to pick up. Led by carriers such as 3, Orange, Vodafone, and T-Mobile, broadband and wideband data services are being launched across the board with favorable adoption.

North America

The American market was a slow starter with data services. Though carriers tinkered around with Cellular Digital Packet Data (CDPD) in the 1990s that offered 15 Kbps, the service never took off. Even TXT took time to be adapted to the level of usage in Europe and Asia. It was Sprint among the top operators in North America that believed in data and launched several new data services in the United States. From the early 2000s, it led the carriers in data ARPU and data's contribution to the bottom line. It is only in the last few quarters, starting in 2005, that Verizon and AT&T Wireless* have started to catch up and eventually pass Sprint. For the U.S. market, 2005, 2006, and 2007 have been high growth years, which will reach 83% subscriber penetration by the time this book is published. It remains the number one market in terms of service revenue generated.[†] In fact, in 1H 2007, it narrowly edged past Japan as the number one market in terms of service data revenues generated and is likely to stay perched at this position for the foreseeable future (Fig. 3.11).

Data services have also experience good growth in the last three years and it shows in the financials. The United States is the only country with three carriers generating more than a billion dollars in data revenues per quarter (AT&T Wireless, Verizon Wireless, and Sprint-Nextel). As of 2007, over 50% of the data revenues were coming from non-TXT applications. Both consumer and enterprise segments are showing significant growth. The data revenues at the top four carriers were growing approximately 20% quarter over quarter in 2007. New services like mobile TV, advertising, and search are being introduced and refined at a fast pace. Carriers are becoming more open to experimenting with new data services. Management has taken notice and are actively investing in future infrastructure in search of double-digit revenue growth contributions from data services (Fig. 3.12).

The biggest impediments to explosive growth, similar to what was experienced in East Asia, are the perceived inequity in revenue share between content and application owners and the closed gardens of the cellular operators. However, the circumstances vary with each carrier, off-portal revenues and market share is

---

* The old Cingular Wireless at the time. Cingular got acquired by SBC, which promptly renamed the company to AT&T Wireless.

[†] In fact, the next four markets of China, India, Russia, and Japan combined generate less revenue than the United States.

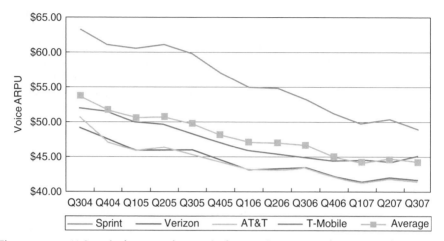

Figure 3.11. U.S. wireless market: wireless voice ARPU (2004–2007). *Source*: Chetan Sharma Consulting, 2008.

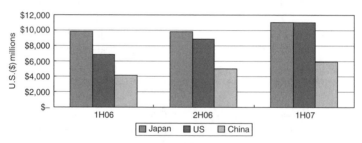

Figure 3.12. Mobile data revenues in top three countries. *Source*: Chetan Sharma Consulting, Global Wireless Data Update 2007, March 2008.

increasing gradually for applications.* With the advent of faster networks[†] and low cost smartphones, it will be difficult for carriers to continue to choke out open subscriber–owned devices from accessing non-carrier-hosted applications. In fact, how the system operators embrace this change of business model will define their long-term success.

In addition to the major markets discussed above, the next tier of markets, which will also experience solid growth, include Brazil, Indonesia, Russia, Argentina, Ukraine, and Africa.

---

*Openness is clearly subjective. Some of the carriers have been guilty of egregious offenses such as disabling Bluetooth from the device or taking out WiFi functionality for the U.S. version of the device.

[†]The United States is actually at pace with Europe on 3G deployments and adoption (except Italy, which has surged ahead with faster adoption).

THE DEVELOPING WORLD IS CATCHING UP RAPIDLY

There is another significant aspect to the global expansion of wireless networks that is commonly found in the developing world. In markets where there was insufficient or nonexistent legacy copper telephone infrastructure, cellular has rapidly increased access to telephone and data services for public that have historically been disenfranchised from participation in the global network. In lesser-developed nations such as Bangladesh, Telenor's Grameenphone has become the dominant voice and data service provider, far eclipsing the historical capacity and capabilities of the legacy monopoly telephone company. Grameenphone has added over 10 million subscribers, mostly in just the last three years, which is far more than the 1 million wireline phones that were previously available to serve this country of over 150 million people. In addition to Grameenphone, there are several other cellular operators including City Cell, Aktel, Banglaphone, and Warid. Combined, these organizations account for approximately an additional 5 million subscribers (circa May 2007). Similarly, in India, Latin America, South America, Africa, and Southeast Asia, the adoption of cellular services as the primary means of voice communications has demonstrated the advantages inherent in large-scale wireless cellular telephone network deployments. Even with ARPU per month of as little as US$5, the operators in these markets almost universally continue to enjoy success and find profitable operations within reasonable time frames. In addition to basic voice services, which in many areas are providing the first telephone calls the majority of the local population has ever made, the inclusion of SMS and enhanced data services is adding tremendously to local productivity and quickly improving their ability to compete for business (Fig. 3.13).

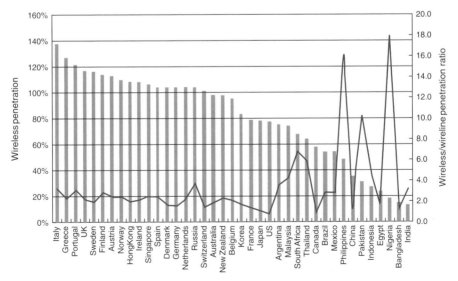

Figure 3.13. Telecom penetration in major countries. *Source*: Chetan Sharma Consulting, 2008.

Bringing the developing world into the era of the broadband Internet should be one of the highest priorities of governments and international aid organizations. Given the already installed cellular infrastructure that most often includes radio transmission towers, generators, and environmentally secure enclosures for electronics, it should now be possible to rapidly deploy next-generation broadband wireless access technologies such as WiFi, fixed WiMAX, and mobile WiMAX as overlay networks, complementing the existing voice cellular infrastructure. Alternatively, in many jurisdictions, it will be inevitable that new service providers emerge to compete with the existing voice-centric cellular operators having pure IP-centric hybrid networks. If the cellular operators fail to address the opportunity to upgrade their existing infrastructure to participate in the wireless IP broadband era, then they will surely face substantial competition. One of the truisms of business is that excess profits breed ruinous competition, and wealthy voice-centric cellular operators who do not anticipate the change of sea state that is upon our industry will surely suffer for the oversight.

REFERENCES

1. Wireless Unbound. McKinsey & Company; 2006.
2. Global Wireless Data Update 2007. Chetan Sharma Consulting; March 2008.
3. In-stat. CTIA Conference; 2007.
4. IFPI. Digital Music Report; 2007.

4

THE VIRTUAL DISPLACES THE PHYSICAL

We are heading into a time when the provision of business and consumer broadband services is proliferating over virtually every type of media, including fiber, coaxial cable, copper, and wireless to both mobile and fixed locations. Wireless will continue its expansion into new broadband applications. Widespread user adoption will be driven by the inclusion of always-on broadband services over all network types in common use, whether they be based on legacy infrastructures, such as copper, coaxial, and broadcasting networks, or on new fiber- and wireless-based broadband infrastructures. The recent frenetic push into Internet Protocol television (IPTV) services by virtually every segment of the telecommunications industry is driving the distribution of image and video content into devices representing each of the three screens in widespread use: the television, the personal computer, and the Internet-enabled mobile device* (IEMD) handsets. Thus, the promise of the broadband future is finally at hand after decades of progress to translate the vision of the convergence of all types of media into actionable service delivery networks, making all forms of content available on any platform at any location.

Nicholas Negroponte developed the concept of the "Negroponte shift" to describe how services that were traditionally delivered over wires (voice) were now shifting to delivery over wireless networks. He noted that services such as television were now being primarily delivered over coaxial cable instead of being

---

* A term authored by Trond Johannessen, CEO, Blackwolf Consulting.

Wireless Broadband. By Vern Fotheringham and Chetan Sharma
Copyright © 2008 the Institute of Electrical and Electronics Engineering, Inc.

transmitted from television transmission towers, while voice communications, which were traditionally transmitted over copper wires, were shifting to wireless cellular networks. It appears that we are now getting ready for a partial reversal of the Negroponte shift, as wireless broadband systems expand the transmission of video to personal mobile handsets, adding a third screen to the traditional choices of television or computer screen for users to access video content. This growing market development might be labeled the "Negroponte reflux." Further fueling this trend is the integration of VoIP onto fiber and coaxial broadband services to the household, pulling some of the cellular voice traffic back into the "wired" environment.

We are struggling through a period of massive change in the basic fabric of our global telecommunications infrastructure. After a century of incremental refinement of the copper wire, narrowband, circuit-switched global telephone network, the past 20 years have been witness to technical change at accelerating rates.

The advent of digital switching and the tremendous capacity increases possible over optical fiber cables drove the final wave of refinement in the hierarchical legacy circuit-switched architectures. Fiber was initially integrated into the network primarily to support long-distance communications, linking centralized switching centers in each city. Under the North American Numbering Plan, local phone numbers are almost universally assigned on the basis of the fixed location of the central office. Calls were connected by switches, which selected the appropriate routing between each endpoint, using dedicated physical elements of the telephone network. This legacy location-dependent architecture is now under widespread assault by the insidious forces of digital progress. Virtual networks routing traffic over packet network facilities shared among multiple service providers are rapidly replacing the hardwired dedicated circuit networks of the legacy telephone system. This revolutionary change creates substantial advantages for users and carriers, while raising a number of challenges, which will be explored in this chapter.

FROM CIRCUITS TO PACKETS

The migration from the dedicated circuit-switched networks to IP-based statistical access networks brings with it numerous new technical and market variables, which will require service providers and their customers to navigate a new landscape of service-level alternatives. These are QoS choices that the circuit-switched networks cannot easily offer as a "one size fits all" architecture.

These will range from low end, simple "best efforts" service offerings to guaranteed high quality delivery of virtual full-period circuits emulated over shared packet network resources. Routed packet networks provide flexibility to create new types of services and to leverage the capacity of common network facilities, regardless of whether they are carried over fiber, wire, or via wireless facilities.

For years our industry debated the merits of traditional time division multiplexing (TDM) versus the first wave of broadband packet core network protocols using the asynchronous transfer mode (ATM) protocol, which ultimately yielded to Ethernet operating under the transmission control protocol (TCP)/IP. For those familiar with Star Trek, it is now safe to say to all contending transmission protocols that "Ethernet is the 'Borg,' resistance is futile, prepare to be assimilated."* This overwhelming dominance is also proving to extend to Synchronous Optical Network (SONET)/Synchronous Digital Hierarchy (SDH) for optical fiber network transmission. Gigabit Ethernet standards are now rivaling SONET in both distance and capacity for core transmission services, with 10 GigE and 100 GigE platforms currently under development. The age of the ubiquitous Internet leveraging end-to-end Ethernet access and transport capabilities is inevitable. The old rules are changing, and as control and choice over previously customer-captive carrier relationships expand across multiple network platforms, the ultimate beneficiary of this major change in how we obtain and disseminate our information and communicate will be the public at large.

"LIES, DAMN LIES AND STATISTICAL ACCESS NETWORKS"

One of the unique challenges facing the emerging wireless broadband industry is how to best define the performance of packet networks to allow customers to differentiate between various services and select from among alternative technologies and vendors. Given the variability in the factors that define network performance, even over the same network infrastructure, it is easy to understand why "maximum data rate" (under what most often appear to be ideal laboratory conditions) has become the metric of choice throughout the industry, despite the fact that this metric is almost always misleading, if not completely useless.

Over the course of the author's career, he has often been asked about the data rate of various packet data radio products with which he has been involved. The challenge of crafting an answer that would satisfy the questioner while also being mindful of the complex nature of the matter has often led him to respond to such inquiries with his own request, "Please ask me a simpler question." Once engaged with the questioner, it was always fun to examine the conundrum by explaining the multivariate maze of issues that complicate the answer. Included among the issues impacting "data rate" are the following:

1. bandwidth
2. power
3. modulation
4. link margin
5. channel configuration
6. contention (overbooking) ratios
7. security
8. management overhead

* CBS Paramount Television.

9. weather considerations

10. interference from both internal and external sources

Our failure as an industry to address this widespread problem head-on with a uniform, industry-wide campaign will increasingly undermine our shared credibility. The famous quote commonly misattributed* to Mark Twain, "Lies, damn lies and statistics," comes to mind, with the thought now applied to statistical access networks, where equal liberties with torturing facts and figures abound in our industry, and these facts and figures are being tossed about by vendors and pundits alike. As an industry, we need to work together to find some "simpler" answers that are based on reality and which share common assumptions. We call upon the Institute of Electrical and Electronics Engineers (IEEE) to inaugurate a "standards setting process for definitions" in search of a common official interpretation of our industry's least precise and most confusing issues—the factual and clearly understandable description of the real world performance of our telecommunications networks and products among a wide range of competing technologies.

MOORE'S LAW FINALLY REACHES TELECOM

Despite our inability as an industry to have thus far developed precise and shared language to describe the new broadband that both the public and practicing professionals can understand with clarity, we are all sharing in the benefits of Moore's law,[†] which is finally extending its use to the telecommunications marketplace. Over the past 100 years, the legacy telephone networks have increased their performance and features largely in a slow, incremental process. Following the growth of the broadband Internet, the influence of the distributed computing environment has been making a significant impact on virtually all aspects of the telecommunications landscape. The increase in performance and the corresponding reduction in costs per unit of computing power have accelerated the pace of change throughout the telecom industry.

Further, we are witnessing the advent of software-defined radio (SDR) systems and cognitive radios, which leverage the capabilities of powerful digital signal processing chips to replace traditional discrete radio system components. Cognitive radio technology is enabling the creation of frequency agile, protocol agile, and application agile radio frequency devices that extract maximum efficiency out of scarce radio spectra. These trends and technologies are examined in detail in Chapter 10. New services and applications based on new products at dramatically reduced expense are at hand, enabling the wireless industry to follow the trajectory that Moore's law had previously set on the computing industry.

Increased power at the edge of networks is also leading to a migration from large centralized switching and core network intelligence to distributed switching,

---

*This familiar statement was actually coined by the famous British politician and Chancellor of the Exchequer Benjamin Disraeli, 1804–1881.

[†] http://www.intel.com/technology/mooreslaw/index.htm

applications, and service platforms that are resident at the network edge. This concept has been accurately described as "ditch the switch."* The distributed intelligence of powerful user devices is what George Gilder has long postulated as the Telecosm.† Gilder's book, *Telecosm*, describes the transition from the traditional narrowband access networks connected to large centralized computing facilities, such as central office telco switches, to a ubiquitous highly distributed broadband backbone network. This telecosm is a high capacity, yet largely dumb network, which links together intelligent computing devices located at the edge of the network and controlled by the users of the network. The telecosm is bound to proliferate because of the ever-increasing capacity of "dumb" broadband transport networks provided by the combination of gigabit Ethernet core networks extended by high speed Ethernet access networks connected to highly intelligent edge devices.

The traditional approach of a smart network supporting dumb customer terminals retained control of the network in the hands of a few large network operators. It has had the same pernicious effect as the "walled gardens" in mobile communications. Because the intelligence resided in the network core, the network operators had to install, operate, and maintain large expensive central switches. This control reduced competition, making it difficult, time consuming, and expensive for competitors to construct competing networks. This situation, in turn, has led to the regulatory "piecing out" of the legacy networks at mandated wholesale charges to permit the entry of competitors and the advent of price and service competition. The housing of the intelligence in the network has also thwarted the creation of a vibrant, innovative customer appliance industry. The customers were largely at the mercy of the network operators in their choice of end-user equipment.

The ensuing reversal of the traditional situs of intelligence away from the network centers and out to the edge has been a great boon to the consumer (as well as to competing network providers). Some examples of products that have emerged as a result of the migration of intelligence to the network edge include the following:

1. the Apple iPod phenomenon
2. personal video and image sharing
3. peer-to-peer computing
4. blogging
5. all forms of edge-caching applications

Gilder's vision has largely been realized in the expansion of the global Internet to include local broadband access capability, whether delivered over fiber-optic lines, coaxial cable, and copper wire via DSL or wireless.

---

* Dr. Rick Baugh, Dr. Heinz Lyklama, and Robert Foster coined this term to describe their pioneering work in distributed routing and switching technology.

† George Gilder's Telecosm: "Telecosm, The World After Bandwidth Abundance, Touchstone, Copyright 2002 by George Gilder"

There is, however, an interesting development that is offsetting, at least for an interim period, Gilder's telecosm. This development is the growth of large-scale server farms and centralized storage and routing facilities, which serve the exponentially growing volume of Internet content hosting, search engines, and address and authentication management. In both the core network and at its periphery, massive amounts of data are being created. These data need to be stored for retrieval and transmission on demand over the public Internet and corporate intranets. At present and at least in the short term, the last-mile access networks typically do not have sufficient bandwidth to deliver these data at the requisite speed. The surging demand for efficient Internet navigation, e-mail, multimedia content, audio and video broadcast, and unicast (long tail) services is thus driving demand for large, centralized hosting facilities and regional caching and storage nearer the network edge. This trend is a by-product of the repurposing of the legacy network backbones from primarily voice- to data-centric services in a world that is still dominated by narrowband first- and last-mile connectivity. Only when massive broadband is extended to the edge of the network will one find a reversal of this trend.

Dramatic reductions in the cost of digital storage, the resurgence of free advertising–supported business models for Internet services, coupled with the distributed computing power capabilities of an ever-increasing Internet Web 2.0* community leveraging user self-generated content, greatly enlarged by the advent of "social networking," are rapidly eclipsing all previous forecasts for Internet storage and transmission capacity. Thus, we find ourselves at an interesting and exciting crossroads, where Moore's law continues to deliver on the promise of powerful, yet low cost computing, meeting the increased awareness throughout the world of the extraordinary value of true broadband access networks that deliver services at affordable prices to users everywhere.

We are going to experience an interesting tension between the growing capabilities of distributed processing and storage and a reconsolidation of centralized facilities, which provide the security and convenience of centrally hosted content and access gateways. Interestingly, this paradox is driven in both dimensions by the assimilation of Moore's law into the broadband wireless telecommunications space. More MIPS (millions of instructions per second) at lower prices will continue to improve performance and reduce the cost of bandwidth throughout the telecommunications industry at all levels.

THE "NEW BROADBAND" ECLIPSES "NEO BROADBAND"

True Broadband, Not Just Relabeled Wideband

As mentioned earlier, our industry is suffering from some serious vagaries with its definitions. In 2000, the FCC relabeled "broadband" to be any bidirectional

---

*Tim O'Reilly has been credited with coining the term "Web 2.0." to describe the world of user community–generated content.

circuit that was greater than 200 Kbps [1]; previously broadband had been defined by the IEEE as a circuit that was greater than 1.544 Mbps duplex. Wideband covered all capacities between 64 Kbps, the upper boundary of narrowband, and 1.544 Mbps. Thus, the "rebranded" broadband adopted by the FCC is in reality only a description of a service that is toward the lower end of traditional wideband. In addition to the confusion these imprecise definitions have created in the marketplace, it has also raised the question of what we should call the rapidly proliferating new services at 10 Mbps, 6 times faster than the IEEE technical definition of broadband at 1.544 Mbps (DS-1/T-1 speeds), or fast Ethernet services at 100 Mbps, which is about 65 times faster than the T-1/DS-1 capacity that defined the original bottom of broadband. We now face the emergence of metropolitan-scale gigabit Ethernet networks operating at about 647 times faster than basic broadband or, more absurdly, 5000 times faster than the FCC's inadequate and confusing definition of broadband at 200 Kbps. Presently, all of these services come under the definition of broadband, which provides no useful information to a public that has been conditioned to believe that services now labeled at the low end of wideband communications are actually broadband. "Just saying it is so, doesn't make it so" [2]. Regardless of what we decide to call these increased levels of bandwidth—"megaband," "maxband," "gigaband," "terraband," or "ultraband," to craft a few possibilities—we need to refine and update our completely obsolete terminology.

What actions can we take to develop understandable and useful metrics of performance for broadband systems and services and to disseminate them throughout society? At a minimum, we suggest this issue warrants convening an IEEE working group to develop and seek comment and socialize a new formal set of definitions, which can be adopted by all organizations that need to position their products and services among a confused populace, including both professionals and the lay public.

REGULATORY AND PUBLIC POLICY COLLIDE WITH TECHNOLOGY SHIFTS

As we enter the next era of telecommunications services, our networks are rapidly migrating away from the telephone networks' historical reliance on geographic determinism. Always-on IP networks challenge many of the traditional approaches to network topology. The location of the IP-terminating device is essentially unlimited. This creates numerous opportunities and challenges for service providers, network access operators, and backbone transport carriers and switching, routing, interconnection, and hosting facilities.

This location independence is especially of concern to enhanced 911 (E911) or emergency notifications, which have historically tied each individual number to a finite location. The FCC has mandated the inclusion of E911 emergency notification services for all cellular companies. It has also established requirements for

VoIP inclusion to provide all users with 911 Public Safety Answering Point (PSAP) connectivity and position location determined to within 300 meters.

We have entered into a time in history where the roles of the vertically integrated and closed networks, owned and operated by carriers, who are also the service providers, are becoming increasingly blurred. New business models are emerging as the Internet expands and improves in all aspects of its ability to support full convergence among all types of services, including voice, data, and video. The attacks on the traditional "walled gardens" of the legacy operators will continue to expand from innovative service providers who will leverage each network element in new ways, including access, transport, and content. As these new business models proliferate, the tension will increase between the legacy providers expanding into new markets to compete with these new types of content and service delivery businesses. The impact that these new business models will have on the revenues of the incumbent service providers will determine whether they decline in influence or remain in their position of industry dominance. The telecommunications business is going to be subjected to change and evolution that will rearrange the market positions of most players in ways that will require thoughtful regulatory policy and business neutral decisions by government regulators and legislators. The list of the contentious issues that will have to be addressed in the coming decade includes such major issues as the following:

1. Historically industrialized nations relied on revenues from traditional long-distance voice termination tariffs to build their networks and subsidize their services to lower income strata and rural subscribers and to incubate new services. The inevitable triumph of VoIP and other Internet-enabled services raises the fundamental issue of how the networks of the future will be financed.

2. A related subissue is whether a subsidy for rural telecommunications services should continue, and, if so, who should pay and who should collect.

3. How will digital rights management (DRM) among all of the various types of content creators, owners, and resellers be addressed? Tension is growing between the legacy providers of mass-market content and the emerging niche market content developers, and over how free market forces will be allowed to determine the future policies and settlement structures for DRM rights holders and users.

4. From where will the replacement of lost tax revenue from traditional telephone services migrating to the largely tax-exempt Internet come?*

5. How will the issues raised by the interception of communications traffic by law enforcement agencies be addressed? This matter has jumped to the forefront of public policy issues at the time of this writing. Congressional

---

*Local telephone sales taxes have historically been one of the largest single contributors to local jurisdictions.

investigations during 2007 revealed that most of the major telecommunications carriers (with the exception of Qwest) had surreptitiously allowed covert government agencies, such as the National Security Agency (NSA), access to their call records on a widespread basis. The Bush administration is fighting to persuade Congress to grant these telecom companies retroactive immunity from prosecution under laws that appear to make the practice illegal. The administration defends the practice as necessary in a time of increased concerns over terrorist attacks and covert operations against the interests of the United States. The administration also argues that traditional policing of these intercepts is too unwieldy and time consuming to allow the government to adequately protect the State and its citizens. Regardless of the outcome of recent policy wrangles, it is likely that the government will continue to intercept communications, with more or less protections for individuals, which will, if the past is prologue, depend on the temper of the times and perceptions of the degree of threats to the well-being of the State and its citizens.

6. What will the impact on the spectrum valuations under auction processes be if the winning bidder must provide open access to the network?

7. Will the advertising-supported network models prove to be sustainable? Will their operators figure out how to share revenues fairly with the "open network" operator who delivers the information to the intended recipient?

8. What policy positions can be established to prevent the incumbent wireless operators from outbidding new competitors due to their unique financial power and then warehousing the new spectrum, given there are few substantive and timely implementation requirements under current regulations?

9. The concept of local franchises granted by municipalities for the delivery of cable television services is under pressure from telephone companies seeking to deliver video services to consumers. Telcos are pushing hard for statewide licensing as a means for them to provide local television services in competition with cable systems, without the time and expense needed to obtain separate franchises from each governmental jurisdiction within a state. This development raises several related issues. Will the telcos be successful in their efforts? Are local wired video franchises sustainable in the era of the Internet and widespread mobility? What mechanism would replace cable television franchise revenues to the local community? What would happen to other traditional provisions of local franchises, such as mandatory local access channels?

10. How will broadband ISPs be treated when they are capable of spoofing cable systems, but with virtually unlimited "channels" of on-demand unicast content? Will this technology shift make obsolete the legacy concept of the local "cable franchise" or the emerging telephone company–delivered equivalents of prepackaged suites of content?

11. The migration to fixed-mobile convergence is inevitable and accelerating. Each traditional type of service provider, such as telephone companies, cable companies, cellular companies, and satellite television broadcasters, is seeking to incorporate the capabilities of others' legacy specialities into fully converged unified networks, serving customers regardless of the environment they may be in at any given moment. This convergence overlap will challenge and disrupt many of the existing regulatory environments and traditional business models across these markets. Some of the issues that are raised include the following, in addition to the ones suggested above such as who should pay what subsidies: (1) What is the role of traditional regulatory oversight of intercarrier settlements for international interconnection and revenue sharing? (2) What regulatory authority should the states and local jurisdictions have? (3) How can the cable multiple system owners compete against the ILECs who can offer a quadruple play, including seamless mobile integration? (4) What regulatory oversight is appropriate to maintain E911 services, which apply to or are even feasible for "second-line" services that are provisioned over either a licensed or license-exempt radio spectrum, but which compete with legacy and common carrier services? (5) What types of handsets and user devices will emerge?

12. Regulatory flexibility will become essential as the blurring of boundaries between the various services that have historically been regulated within distinct vertical silos evolves. Local telephone, Intra-LATA toll (regional long distance), interstate long distance, international long distance, terrestrial television, satellite and radio broadcasting, media cross-ownership, Internet access, cellular telephony delivered over new packet data architectures, and a whole host of tangential regulatory and taxation treatments must be developed and implemented in a highly fractious and politically charged environment.

The answer to these and related multivariate questions is central to ensuring a successful transition of the traditional telecommunications industry into the broadband IP environment of the future. There are a growing number of "900-pound gorillas" showing up in the room, and if we fail to address the implications of these technical and market-changing events in a thoughtful and comprehensive manner at all levels of the political, regulatory, and financial communities, it will certainly lead to disruption and pain across all factions contending for participation in the broadband future. The following chapters will attempt to identify the key issues that will influence this transition and posit some of the major trends that will emerge through the process to affect the participants and the public at large. We are at a major crossroads in the evolution of our information and communications industry. We will enjoy a digitally empowered future only if all the key decision makers and constituencies cooperate in a manner that will provide a future regulatory framework that does not constrain innovation and progress simply to protect rapidly obsolescing legacy technologies and networks.

REFERENCES

1. Local Competition and Broadband Reporting. CC Docket No. 99-301. Report and Order, 15 FCC Rcd 7717, 7731, 2000.
2. Festinger L. A Theory of Cognitive Dissonance. Stanford: Stanford University Press; 1957.

5

CONVERGENCE FINALLY ARRIVES

The realization of the decades-long vision of a fully converged telecommunications network and services environment is being enabled by the expansion of fiber- and coaxial-based broadband access networks, coupled with the progress in the wireless sector to provide true broadband capabilities to individual user devices. The hybrid synergies between the proliferating broadband fixed networks and the emergent broadband wireless mobility networks will redraft all of the traditional lines between and among the network operators, service providers, and subscribers. Ubiquitous broadband access to all services and all forms of content regardless of location, time, or media will be realized within the coming decade. How we as a society leverage this remarkable and fundamental shift of capabilities is a nontrivial challenge that will reach into every community, industry, organization, and family.

THE QUAD PLAY: VOICE, DATA, VIDEO, AND MOBILITY

The invasion of formerly discrete franchise areas by competing service providers seeking to enhance their revenue potential is rapidly proliferating. We now have telephone companies moving into video distribution and cable companies seeking to provide voice services and high speed Internet connectivity. In parallel with these developments, the cellular operators are seeking to improve their offerings to include video services, both on-demand unicast and broadcast reception, while simultaneously improving their mobile data capabilities. Moving through all of

Wireless Broadband. By Vern Fotheringham and Chetan Sharma
Copyright © 2008 the Institute of Electrical and Electronics Engineering, Inc.

these new integrated service platforms is the customer-driven trend for simplification and unification of service provider relationships. The demand for obtaining a full range of services from a single supplier is driving the industry to full convergence, compelling each sector player to seek ways to add the missing elements it needs in its portfolio of services to address each of the key domains for the quad play: voice, data (Internet access), video, and mobility. Although wireless can play a significant role in each domain required in the quad play, it is with mobility that the utility value of radio reigns supreme, regardless of whether it is untethered communications in the home or office, metropolitan-scale personal locations, or high speed communications in vehicles.

The cable companies, the DBS service providers, and the major Internet portal operators are deeply disadvantaged at the present time in seeking to deliver the full suite of quad-play services. The cable television industry had earlier taken a first step toward its full participation in the quad-play wireless service in addition to its core franchises. Four of the leading cable industry companies (Comcast Corp., Time Warner Cable, Cox Communications Inc., and Bright House Networks and their cellular implementation partner Sprint Nextel Corp.) organized a consortium, the SpectrumCo LLC coalition, which acquired AWS spectrum licenses covering approximately 90% of the U.S. market. The SpectrumCo licenses were virtually all in the B block of the AWS spectrum, providing 20 MHz of paired 3G spectrum. Sprint withdrew from this consortium in August 2007, and at the time of this writing, there have been no public announcements regarding how these assets will be utilized, although they have entered into the new Clearwire consortium.

Several new national wireless networks are in the advanced planning stage at present, and the first large-scale mobile WiMAX deployments by Clearwire is in the process of deployment. In addition, the FCC has completed the 700 MHz auction, with AT&T and Verizon capturing the majority of the national licenses. This long-pending auction attracted the interest of virtually all of the major players in the wireless industry, and under the pressure exerted by Google, the FCC agreed to incorporate the concept of open systems into at least one of the three spectrum block licenses up for purchase. The addition of new participants onto the national wireless service provider stage was hoped to accelerate the evolution of the legacy cellular operators into broadband wireless 4G systems earlier than they would have appreciated if left to their own vision of "walled gardens" and extremely sticky customer relationships. The very concept of real-time user control over selection of service providers is alien to all of the business cases of the established cellular operators in the U.S. market at the present time, and with the exception of Echostar's winning bids for a significant national footprint, the bulk of the 700 MHz licenses went to Verizon and AT&T, thus ensuring their continued hegemony over the mobile wireless market.

THE QUAD-PLAY ADVANTAGE

Bundle awareness is growing. Consumers are increasingly comfortable with nontraditional providers. In a recent survey by Yankee Group, 16% of consumers

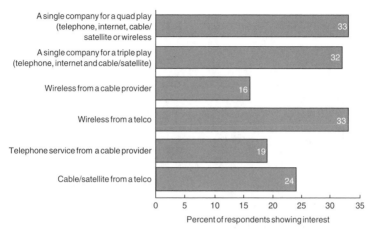

Figure 5.1. U.S. consumer attitudes on bundle offerings. *Source*: Yankee Group.

would likely purchase wireless service from a cable company and 24% would likely purchase satellite from a telco. These data show that bundle momentum is building, and the consumer is becoming knowledgeable about the bundle and provider options (see Fig. 5.1).

- In 2005, 47% of consumers were interested in using a single service provider for most or all of their household communications services.
- Consumers are interested in a single bill and a discount. Eighty-one percent of consumers are interested in a single service provider because of the benefits of receiving a single bill. Sixty-eight percent are interested because of a discount.
- Although consumer readiness for the bundle is strong, adoption rates especially for triple- and quad-play bundles are still low.
- Approximately 35% of all households currently have a bundle. The majority of these are double-play bundles.
- Although one-third of consumers are interested in triple- and quad-play bundles, only 3–4% of households have a triple- or quad-play bundle.
- Today, 27% of the U.S. households subscribe to all three products in the triple-play bundle: broadband, paid television (cable or satellite), and local phone service. With only 3–4% in a triple-play bundle, this leaves a current market opportunity of 24% of the U.S. households (see Figs. 5.2 and 5.3).

Reasons for Quad Play

The benefits of the bundle are lower churn, increased ARPU, and higher customer satisfaction ratings. The combination of these benefits creates a desirable customer, a customer who is likely to be more loyal than the average customer. Triple-play-services-bundled offerings of TV, phone, and Internet have proven

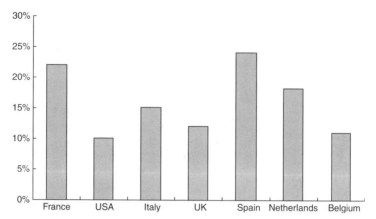

Figure 5.2. Triple-play (Internet, voice, video) penetration. *Source*: Pyramid Research.

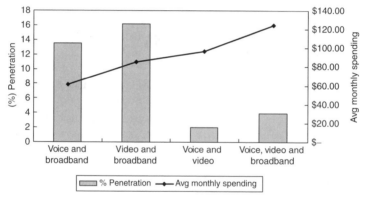

Figure 5.3. Double- and triple-play penetration in the United States. *Source*: Yankee Group.

effective for operators and service providers in reducing customer churn and delivering operational economies. Some operators believe that adding a fourth element, wireless, will improve the equation still further.

Network consolidation for the sake of triple- and quad-play services can reduce operational costs, but brings new challenges as well. Its three layers of management—traffic control, equipment control, and administration/billing—all have to work together and be interoperable, which means that some standards will need to be developed.

Other challenges include QoS issues and network "policy control." Some operators expect to be compensated for the extra bandwidth required to support hosted VoIP services run by their partners. Billing is complicated in these multi-mode arrangements too, and there are questions of privacy: to prioritize network

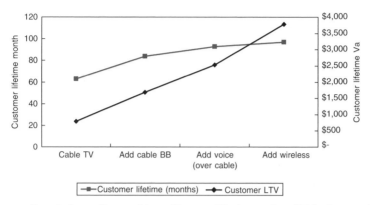

Figure 5.4. Quad play: effects of bundling on lifetime value (LTV). *Source*: inCode.

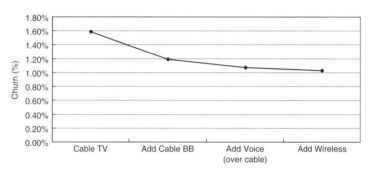

Figure 5.5. Quad play: effects of bundling on churn. *Source*: inCode.

traffic, the IP packets must be examined in some detail, identifying what individuals are doing online (see Figs. 5.4 and 5.5).

Bundle Offers

There are different permutations and combinations available for cable/MSOs, telcos, and wireless operators (see Fig. 5.6).

As markets are becoming saturated, quad play has become an important strategy for telecom service providers to bundle services to reduce cost to the subscribers and increase customer loyalty and thus decrease churn. In the coming days, almost all of the major players will launch some form of quad play either on their own or through their partners.

FIXED MOBILE CONVERGENCE AND UNLICENSED MOBILE ACCESS

Cellular operators have been seeking ways to leverage their mobile network capabilities into the home, the traditional domain of the fixed-line telephone

Service Provider	Missing component	Strategic Options	
		Partnership	*MVNO*
Cable (Offering Voice, Broadband Internet, and Video)	Wireless	Sprint-Nextel and Cable companies	NTL with Virgin Mobile
		Partnership	*Integrated Ownership*
Wireless (Voice, Broadband wireless Internet, and Wireless)	Video	Sprint-Nextel and Cable companies	None
		Partnership	*Integrated Ownership*
Wireline (Voice, Broadband Internet)	Video	SBC and Dish Network	France Telecom
		Integrated Ownership	*MVNO*
	Wireless	Bellsouth WiMax trials	BT Fusion

Figure 5.6. Different strategic options available for service providers.

companies. With the advent of WiFi access points that can be easily and inexpensively connected to subscribers' own DSL or cable modem high speed Internet access networks, a new hybrid platform for fixed-mobile convergence has arrived. Both T-Mobile* and Sprint[†] are marketing dual-mode handsets that function either over their home WiFi (T-Mobile) or Fempto cell (Sprint) access points or on their respective mobile networks when subscribers are away from home.

The T-Mobile solution also supports VoIP access over its public access WiFi hot spots. Integrating a VoIP solution for low cost calling from home or participating public access points adds two positive aspects for cellular operators. The Hotspot@Home service charges a $10 per month fee in addition to any standard cellular billing plan. T-Mobile is selling either D-Link or Cisco Linksys 802.11(g) hot spots for $49.95 and a dual-mode GSM/WiFi handset to enable the service. The ability to leverage the cellular network features, and to make unlimited long-distance calls within the U.S. coverage area of T-Mobile, and the utilization of the mobile number into a home-fixed implementation drive the presence of the mobile operator directly into competition with the legacy telephone companies for home services. The service also provides a reliable addition to ARPU while further increasing the stickiness of the carrier relationship with the subscriber and thus reducing churn.

In parallel, Sprint is trialing the implementation of its Airave Fempto cells in selected markets that leverage a customer-owned broadband Internet connection to support a cellular technology (compared with WiFi) private home base station

---

* T-Mobile Hotspot@Home.
[†] Sprint Airave CDMA Fempto cells.

that allows subscribers to continue using their cell phones while at home, without connecting to the Sprint network cell towers, thus saving the subscribers' airtime charges while still delivering access to all of their usual cellular service features for a flat monthly fee. These services are aimed at displacing the traditional landline plain old telephone service (POTS) by leveraging broadband Internet backhaul (provided by the subscriber under "bring your own broadband" approaches) into cellular operator–owned soft switches. The service allows subscribers to talk with no limits while in their homes, without consuming airtime minutes. The program covers unlimited incoming and outgoing calls and nationwide long-distance services. The Sprint Fempto service is unique in that it does not require the user to acquire a dual-mode (cellular/WiFi) handset (see Fig. 5.7). Instead, it simply supports a connection to the users' existing CDMA cellular handset. The Airave services are priced in addition to existing cellular plans at $15 per month for single-number subscribers and at $30 per month for a family plan. Figure 5.8 depicts the Sprint Fempto access point supplied by Samsung.

The Three-Screen Imperative

A key element of the convergence phenomena is the need for video content to be seamlessly transportable between and among the three screens that most consumers use regularly: the traditional television screen; the computer screen, both

Figure 5.7. Fempto cell network architecture.

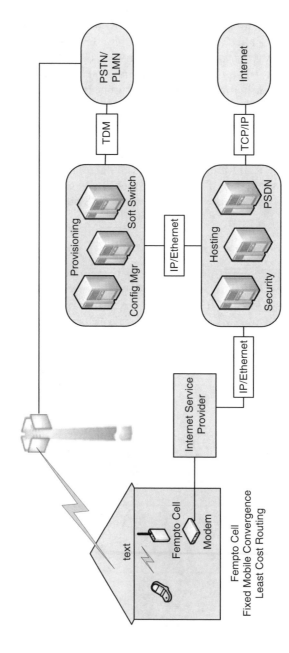

Figure 5.8. Samsung 1900-MHz Ubicell (CDMA/WCDMA) Fempto cell.

desktop and laptop; and their personal mobile devices. Fully converged service providers are going to be seeking networks and technologies that allow them to provide authenticated services to their customers over each of these environments, and oftentimes, simultaneously.

The industry is faced with a new set of requirements for billing and operational support services to control digital rights management, audit trails, and integration with all of the legacy services delivery platforms and middleware already central to running business.

Increased Bandwidth to the Premise/Fiber and ADSL2

Fiber to the home (FTTH) and fiber to the "X" (e.g., node, premise, curb, business) will be joined by improvements in the copper plant with the implementation of ADSL2, improvements to the digital cable plant improving the performance of coax and fiber hybrids, and the advent of metropolitan wireless broadband networks providing broadband services up to 10 Mbps to personal mobile devices and fixed wireless solutions ranging from 10 Mbps up to multiple gigabits per second.

Gigabit Ethernet services are now available even over passive optical systems (GE-PONs) that are providing service delivery of 100 Mbps per drop. The cost per home served is approximately $2400 for buried installation and $1500 for the average overhead installation of GE-PONs. The cost per home served by all cabled media must also account for the "cost per home passed" as the neighborhood cabling must also be included in the financial expense of service deployment. Coaxial cable systems are averaging about $2400 per home passed. These provide a baseline for comparison of the economic utility value of broadband wireless infrastructure of either equivalent or proportional capacity.

Fiber-to-the-Premise Deployments

Verizon FiOS. On the basis of recent press reports, Verizon has reached over 7 million homes passed with its fiber-to-the-home service and had signed up in excess of 943,000 customers by the end of 2007 for the service. This early penetration rate of only about 10% bears a heavy burden of capital expense of approximately $7650 per subscriber of stranded (fallow) network assets. The true test of its FTTP initiative will be in how successful it is at converting a much greater portion of the homes passed into subscribers. Verizon expects to have at least 6 million homes passed by year-end 2007.

Verizon has announced that it anticipates investing approximately $18 billion by 2010 to extend its FiOS network to capture 7 million FiOS Internet customers and up to 4 million FiOS TV customers by the end of 2010. This equates to approximately $1636 per subscriber in capital expense.

In contrast to the industry averages quoted above for the cost per home passed, Verizon is claiming that its capital expenditures have declined to $873 in August 2007 per home passed, with a 2008 target of $850. The expense to connect is presently running at an average of $933. These expense figures were released by the company with only limited visibility into the percentage of homes passed actually converting into subscribers, resulting in an obfuscated view of a head-to-head comparison with the cable industry's metrics.

AT&T U-Verse. As of early September 2007, AT&T is claiming in excess of 231,000 subscribers for its high speed Internet and IPTV service called U-verse by the end of 2007. These subscribers are pulled from the approximately 5 million "living units" that they claim to presently pass for an approximately 2% penetration of their initial installed network capacity. The AT&T business challenge will be to substantially increase its penetration ratio to offset the stranded capital in the 98% of the network facilities not subscribed to by residents of the homes passed.

In all fairness, it is still very early for the telephone companies' foray into the television and true broadband Internet business, but these early penetration figures are well below the historical early subscriber conversion among the cable TV operators. Given the presence of a well-established and mature market operation by the cable operators in virtually every location, the telephone FTTP projects are rolling out; it will ultimately be a marketing war between the new fiber-based operators and the existing coaxial and hybrid fiber coax (HFC) networks to acquire (or retain) sufficient customers to justify their business model. The nuance of claiming "living units" versus homes passed implies a heavy reliance on deployments in high density, multitenant areas for their initial rollouts. AT&T anticipates reaching over 8 million living units passed by year-end 2007.

Hybrid Networks/Stratified Networks

Coaxial Cable: Digital Convergence and Fiber Hybrid Coax. The larger MSO cable TV companies have been moving aggressively into the triple-play space for integrated video, high speed Internet, and voice telephone services to both improve their operating margins and compete successfully against the newly aggressive competitors from the telephone companies, especially Verizon and the reconstructed AT&T. The cable companies, however, are still largely disenfranchised from full participation in the fourth domain of convergence, which is wireless mobile services. How they will respond to this gap between themselves and the telephone companies will be immensely important to the next wave of competition.

COMCAST. The cable industry has been making substantial gains in the deployment of triple-play services with about 3.1 million Comcast Digital Voice subscribers signed up, making Comcast the largest VoIP service provider in the United States. Thus, approximately 12.8% of the Comcast basic video customers are now Comcast Digital Video subscribers, leaving substantial room for growth among their existing 54.2 million subscribers. Penetration of Comcast's existing customer base of high speed Internet service is in excess of 26% at 12.4 million.

Comcast presently passes over 47 million homes, with 24 million basic cable subscribers, resulting in about 51% penetration of its homes passed footprint.

TIME WARNER CABLE (TWC). The footprint of TWC passes approximately 27.6 million homes passed. The company enjoys approximately 14.4 million basic subscribers within this service area, which represents a substantial penetration rate of approximately 52% at a point of relative market maturity, compared with the early stage penetration of the FTTP initiatives by Verizon and AT&T.

CABLEVISION SYSTEMS CORP. Serving approximately 4 million subscribers in the greater New York region, the company passes more than 4.5 million households and 600,000 businesses. It offers the full triple-play complement of video, data, and voice services.

COX COMMUNICATIONS. Cox is the fourth largest cable operator in the United States, with approximately 6.5 million subscribers. It provides video, high speed data, and voice telephone services to its customer base.

The cable companies are at a disadvantage to AT&T and Verizon because of their lack of large-scale wireless services. This leaves them out of the bidding for the quad-play bundles that customers are seeking from the restored telcos reunited with their cellular operations.

Copper – ADSL2, VDSL. In the United States the domain of DSL technology in its many enhanced forms is now firmly ensconced within the legacy telephone carriers. The failure of the Telecommunications Act of 1996 to provide sustainable open and fair access to unbundled network elements of the legacy copper telephone networks can now be clearly seen as a failed policy initiative. In other nations, such as the United Kingdom, Australia and soon New Zealand, a full separation of the legacy telephone company monopolies into separate, arm's length entities has succeeded extremely well in bringing substantive competition for telecom services into reality. In these markets, the legacy monopoly operator was split into a network facilities operator selling rate regulated access and transport to all comers on an equal basis, a wholesale division, and a retail operation purchasing its underlying access and transport from the facilities provider and wholesale operator on equal terms to all competitors. Under this new approach, true competition should be able to emerge, not just the pseudo competition that was proven largely impractical in building sustainable value for the shareholders of the new competitors who emerged to challenge the ILECs post the Telecom Act.

Recently, there has been significant progress with the concept of bundling copper into sufficient pairs to support metropolitan Ethernet services over copper. Hatteras Networks has pioneered this copper pair bonding technology, and recently both ILECs and CLECs have been facing significant challenges in defining the most cost-effective means of delivering metro networks that previously could only be served by dedicated fiber or microwave wireless facilities.

Real-world experience has demonstrated a 6,000 feet distance average limit to reliably deliver full native speed (10 Mbps) Ethernet circuits. This distance should be an interesting metric for the planners of wireless broadband networks, as they seek to differentiate their service offerings from the fixed networks.

BROADBAND MARKET OVERVIEW

Multiplatform battles heated up in the year 2006, with companies in the industry using every means at their disposal to steal customers from the competition and working hard to keep the ones they had. During 2006, a number of cable operators turned around years of losses on basic video services and added record numbers to their digital platforms with advanced offerings like video on demand (VOD), digital video recorders (DVD), and high definition TV (HDTV). According to publicly reported subscriber numbers, cable MSOs added more than 5.3 million digital video customers in 2006. Comcast, TWC, Cablevision, Insight, and others also reported gains for their analog video products (see Fig. 5.9).

DIRECTV and EchoStar's DISH networks are the top providers of pay-TV. Satellite radio offerings from XM Radio and Sirius continue to attract consumer attention as millions of new cars enter the fleet equipped with satellite radio receivers installed at the factory.

Satellite TV had good subscriber gains in 2006. At DIRECTV, fourth-quarter gross subscriber additions totaled 1.021 million, an increase of 6% when compared with the fourth quarter of 2005, so that DIRECTV added 275,000 new subscribers in 2006. DIRECTV said its focus on attaining higher quality subscribers combined with the significant increase in subscribers with HD and DVD services were major contributors to a reduction in monthly churn.

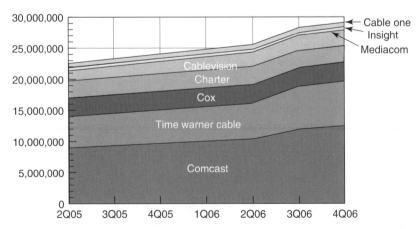

Figure 5.9. Digital cable growth 2Q05–4Q06. *Source*: Company Reports (2007), The BRIDGE 2007.

DIRECTV has concentrated on HD/DVD/VOD offerings, arguing that video is the true killer application for consumers. In early 2007, DIRECTV was offering local HD programming in 57 cities, representing more than 67% of the U.S. TV households, and is promising local HD service in up to 75 markets plus up to 100 national HD channels by the end of 2007.

At EchoStar, the company's DISH network added 350,000 net new subscribers during the fourth quarter of 2006, giving a total of about 13.105 million

TABLE 5.1. Subscription TV Services Subscribers by Provider End 2006[a]

	Service provider	Period	4Q 2006
1	Comcast		24,161,000
	Comcast Digital		12,666,000
2	DirecTV		15,950,000
3	Time Warner		13,402,000
	TW Digital		7,270,000
4	DISH Network		13,105,000
5	Cox		5,430,000
	Cox Digital	3Q 2006	2,700,000
6	Charter		5,433,300
	Charter Digital		2,808,400
7	Cablevision		3,127,000
	Cablevision Digital		2,447,000
8	Brighthouse	2Q 2006	2,275,000
9	Mediacom		1,380,000
	Mediacom Digital		528,000
10	Insight		1,322,800
	Insight Digital		621,600
11	Suddenlink	2Q 2006	1,269,000
12	Cable One		693,550
	Cable One Digital		213,873
13	RCN		371,000
14	WideOpenWest		361,200
15	Bresnan		293,500
16	Service Electric		287,800
17	Atlantic		265,164
18	Verizon FIOS TV		207,000
19	Midcontinent	3Q 2006	196,313
20	Pencor Services	3Q 2006	183,700
21	Knology		178,618
22	New Wave	3Q 2006	164,400
23	Millennium	3Q 2006	157,800
24	Northland Cable	3Q 2006	146,940
25	Buckeye	3Q 2006	146,446

Note that MSOs have their basic subscriber count first and digital subscriber numbers second if applicable.

[a]The Bridge, March 27, 2007.

TABLE 5.2. Cable ISP Subscribers

	Cable ISP subscribers	4Q 2007
1	Comcast	12,900,000
2	Time Warner	7,700,000
3	Cox	3,600,000
4	Charter	2,600,000
5	Cablevision	2,200,000
6	Insight	720,000
7	Mediacom	640,000
8	Cable One	289,010

TABLE 5.3. Dedicated ISP Subscribers

	Generic ISP subscribers	4Q 2007
1	AOL	10,100,000
2	EarthLink	4,200,000
3	United Online	1,900,000
4	Clearwire	310,000
5	Hughes DirecWay	260,000
6	Localnet	260,000
7	First Communications	240,000

TABLE 5.4. Telephone Company ISP Subscribers

	Telco ISP subscribers	4Q 2007
1	AT&T (Including SBC and Bell South)	17,900,000
2	Verizon	8,000,000
3	Qwest	2,500,000
4	Embarq	1,200,000
5	Windstream (Alltel and Valor)	830,000
6	CenturyTel	530,000
7	Covad	510,000
8	Citizens	500,000
9	Cincinatti Bell	220,000

subscribers at the end of 2006, an increase of 1.065 million subscribers for the year. DISH also has HD/VOD/DVD services. In early 2007, DISH offered about 30 national HD channels and promised to bring local HD services to around 50 markets by the end of 2007. The company has a VOD service that is actually an off-hours download-to-DVD service. DISH and DIRECTV, like virtually all other multiplatform players, have consistently subsidized equipment and installation costs for their customers (see Table 5.1).

While these tactics have helped buoy DBS growth, both satellite TV service providers are looking for bundled solutions to sell. Both DIRECTV and DISH have video/voice/data deals offered via partnerships with telcos, yet are still disenfranchised from participation in mobile service offerings.

Internet Access Service Providers

In 2006, there was continuing growth for cable modem, DSL, and other broadband services going to the home, and the nation's broadband total topped more than 53 million in 2006 (Leichtman Research Group). According to the data firm, the top cable broadband providers have a 55% share of the market with 29.3 million high speed Internet subscribers. Cable companies added 4.75 million broadband subscribers during the 12-month period. Telcos have 24 million broadband users. The top telcos netted 54% of broadband additions in 2006, adding 5.5 million broadband subscribers for the year.

The Tables below show subscriber counts for cable companies, ISPs, and telcos (see Tables 5.2–5.4).

Satellite: Satellite Terrestrial Dual-Use Spectrum

There is a 500-MHz swath of thus far unused bandwidth that is coprimary with the DBS TV spectrum at 12.2–12.7 GHz. The band is called the Multichannel Video Distribution and Data Service (MVDDS). This is potentially a major national broadband fixed wireless franchise. This spectrum franchise is largely owned by two principal auction winners representing Cablevision and a consortium that appears to represent the interests of a major satellite system operator. Cable operators were prohibited from bidding in their existing service areas.

This spectrum is licensed for use as terrestrial "wireless cable" or adjunct to DBS service provision of services in the same band. The concept of spatial diversity is used to isolate the MVDDS signals from interference with DBS receivers and to protect the receivers from satellite interference. Further enhancements to these systems are anticipated from the use of coding and advanced modulation techniques to further protect both of the coprimary licensees in these frequency bands.

6

DRIVERS OF BROADBAND CONSUMPTION

TRENDS IN MOBILE AND CONVERGED CONTENT MARKETS

This chapter reviews the applications that impact the growth of wireless broadband. The search for the "killer app" has largely been displaced by the notion that the key is bandwidth itself. Broadband availability enables virtually every type of application, thus making the concept of the "long tail" come into reality with the widespread access to literally every type of niche market, specialized content in existence.

USER INTERFACE

There are few areas of intersection between broadband service delivery and the users of the technology that are more important to the user experience than the user interface (UI). Since the beginning of the mass market for computers and the subsequent popularization of mobile communications, the progress in the area of the UI has been torturously slow and the visual landscape and control interactions provided by the UI have seldom advanced beyond the realm of the arbitrary and cryptic. A few generations of users have now been raised who have assimilated the industry's arbitrary default standards for manipulating and interacting with the wireless and computer-based communications devices. Indeed the industry has been the focus of new language and user-created applications developed by the user community, completely outside the design vision and product road maps of the

Wireless Broadband. By Vern Fotheringham and Chetan Sharma
Copyright © 2008 the Institute of Electrical and Electronics Engineering, Inc.

carriers and solution providers. One does not have to look far beyond the global phenomena of instant messaging (IM) or "texting" using the SMS capabilities of cellular phones to gain the humbling insight into the creation of a new language, which is being developed daily by millions of users swapping text messages at a frantic pace.

Much to the horror of English teachers everywhere, this industry has spawned a new dialect that is virtually inscrutable to the uninitiated. For example, the previous sentence in "txt speak" would look like this: "mch 2 d horror of en tchaz evrywhr dis ind. hs spawned a nu lingo, dats virtually inscrutable 2 d uninitiated."* The reader may find it worthwhile to spend some time visiting NetLingo, the Internet dictionary Web site at http://www.netlingo.com/index.cfm, for a comprehensive list of acronyms and shorthand symbols that will be required to keep abreast of the next generation of wireless users. It will be extremely interesting to ascertain the impact that added bandwidth, enabling low-cost real-time image and video communications between individuals and among selected groups, both formal and informal, will have on the shorthand written language of the texting community. Will rich media turn texting into a short-lived by-product of the narrowband era of cellular communications, or will the intimacy and privacy of texting extend into the broadband future? This is a big question for the future product planners to address as our wireless universe continues to expand in all dimensions.

However, a number of promising developments are starting to make a positive forward impact on the UI world. The first and most obvious is the brand-specific customization of the UI by carriers. The addition of the MVNO business model into the service provider sector has resulted in a need to differentiate these new service providers from their upstream carrier suppliers (most often Sprint Nextel). Rather than retaining the factory supplied (and often boring) UI, the MVNOs, such as the former ESPN, Disney, Helio, and Amp'D, fine-tuned and customized the UI on the devices sold to their subscribers. These new UIs looked much better than those supplied by the mainstream carriers. This new attention to using the UI to promote ones brand and to making the interaction between customer and phone more engaging has resulted in an acceleration of the emphasis placed on the UI by the major carriers. The inclusion of Flash memory into the cellular handset is enabling an increase in the ability to have dynamic graphics and customized UIs on user devices.

In the United States, Verizon is committed to Flash-equipped handsets, and other service providers are likely to follow suit. Device manufacturers such as Samsung and Nokia have also struck licensing deals with Macromedia (Adobe) for Flash. Flash has already been very well received in Japan and Korea, which are well beyond the United States in the penetration of 3G cellular services at the present time. In Japan, over 60% of the devices are Flash Lite capable, while presently in Korea, adoption has just started with about 20% of the device base being Flash compatible. Sprint is embracing the uiOne (from Qualcomm) platform, which is an extension to the Binary Runtime Environment for Wireless (BREW) applications enabling solution. uiOne addresses the ease-of-use issues that have previously plagued the customization of the handset UI by carriers and end users alike.

---

* Translation courtesy of the Lingo2Word. Web site: http://lingo2word.com/translate.php.

Qualcomm's uiOne application provides consumers with CDMA-based phones with a simplified set of tools to express their individuality and to easily integrate the value-added services they seek, ranging from the traditional ringtones and wall-papers to a wide range of hosted solutions enabled under the BREW applications engine. Using the uiOne offering, operators, content publishers, applications developers, device manufacturers, and consumers all have the ability to design their own UI experience and ensure that all elements involved, including colors, fonts, sounds, and function, conform to their own ideas, goals, and visions for the UI experience.

Flash implementation and MYDAS (from Openwave) provide a richer means to control the UI not only for traditional screens but also for active screens. Active screens are noteworthy for their ability to support real-time information being delivered to the device display. This is an area of the UI experience upon which device manufacturers are paying a lot of attention. Motorola has introduced SCREEN3 technology that enables a push data stream feed of news or events in short previews, which, if of interest to the user, can be expanded into more detailed summaries, or the entire content by opening a mobile browser to the Internet. Flash is likely to have the biggest impact on the UI market, as the early users are responding with positive reviews, and to support its widespread implementation there are thousands of existing developers who can provide an efficient application development resource able to leverage mature and prolific tools that were created for the PC software development industry.

HANDSET DISPLAY GRAPHICS

Qualcomm's MSM chip sets provide a multimedia accelerator that enhances pictures and video on the handset by supplying additional APIs and processing power of 15 to 20 MHz just for dealing with decoding of multimedia content. TI's OMAP and Philips' Nexperia platforms have similar functionality.

Multimedia accelerators, such as 2D/3D graphics, audio, video, or Java, are hardware-based processors. These are nonprogrammable chips based on either a DSP core or a microprocessor unit (e.g., high speed CPU ARM processors or MIPS [microprocessor without interlocked pipeline stages] RISC [reduced instruction set computer] processors). The function of a hardware accelerator is to deal with processing part or all of a specific application code. Multimedia accelerators can be either integrated into the modem or used as separate chips that can be connected to the modem. Hardware accelerators have the advantage of being low power consuming and low memory (both random access memory [RAM] and read-only memory [ROM]) demanding. They are based on an architecture specifically designed to maximize the performance of a given application (e.g., audio, video, image, or graphic processing). However, these chips are less flexible in their use as their design is based on hardware. Consequently, they are not upgradeable and do not allow applications to be reconfigured. Hardware accelerators are a good solution for cost-sensitive feature phones aiming to a specific segment of the handsets market, such as onboard game phones, camera phones, or music-enabled phones. However, they cannot deal with general-purpose processing required in feature-rich phones and smart phones. These phones are generally designed to support a number of different applications and services, and the key characteristics include application reconfigurability and the flexibility for service upgrade. In addition, as technology and standards change, mobile phones based on hardware accelerators will be increasingly difficult to upgrade.

In this design, the baseband and the applications processor are incorporated as discrete modules, each embedding one or more processing cores. For example, the baseband might include only a DSP core to run all the three layers of the wireless protocol stack (WPS), or it might additionally incorporate an ARM core to handle the layer 2 (medium access control and peripherals) and the layer 3 (application layer) of the WPS, while the DSP deals with the layer 1 (wireless data link and real-time operations) only. Separate applications processor is suitable for smart phones and feature-rich media phones. These devices traditionally come with different media capabilities, including Internet browsing, advanced games, quality video and audio playing, TV, and multitasking. The device dealing with these tasks must be equipped with a powerful applications processor to ensure performance, availability, and processing flexibility for all applications.

As mobile phones are handling increasingly advanced applications and featuring sophisticated functions that support the mobile lifestyles of their owners, the complexity of UI design is increasing. This growing challenge is made more difficult as the pressure to reduce time to market has not diminished. Device vendors need to increase their flexibility to better respond to the rapidly changing mobile handset market and answer the requirements of operators. The UI has become central to the success of next-generation mobile operators' services. Indeed, vendors are facing strong pressure from operators and service providers to adapt the UI with customized features to improve service usability and facilitate the promotion of branded content. The majority of leading operators, including NTT DoCoMo, Vodafone, Orange, T-Mobile, and O2, are increasingly involved in handset software design. They currently dictate their recipes to both software developers and device vendors. Traditionally, the UI has been implemented as a static layer over the OS. This means that the UI code is an embedded component, making that UI specific to a handset with no upgrades or modifications possible. This approach makes it hard for manufacturers to quickly respond to market changes, adapting the UI to the users' needs for personalization or to tailor it to evolving customization requirements of different operators (Fig. 6.1).

Figure 6.1. Flash-capable Symbian handsets from Nokia.

As the demand for functionality and Internet access and browsing continues to increase among the mobile-user community, the UI will be a fertile ground for competitors in all corners of the industry to add value, differentiate their service or content, and improve the intuitive aspects of user interaction with their devices. The race is on for the developers and inventors to apply their creativity to this ease-of-use opportunity and the impact their enhancements will have on our lives.

MOBILE VIDEO CONTENT

Mobile TV, and more particularly, mobile broadcast video, is one of the most talked about new services for wireless subscribers to emerge in sometime. The reasons for this attention are apparent. Everyone can grasp the concept of mobile broadcast video and immediately relate to it. The migration to fully converged networks, allowing users to have access to all of the trusted and desired services regardless of where one is, is a driver of this next leap forward in the wireless industry. The opportunities are enormous for the network operators, their wireless carrier partners, and the content owners, who see an entirely new channel through which they can reach new customers (and maintain their connection persistently with existing ones).

Mobile video also has some substantial business risks associated with it, including the high costs of building mobile broadcast networks. Mobile broadcast networks are attractive to wireless operators because they do not have to use the scarce capacity of their extremely valuable cellular airtime networks to transmit video to their customers. Video files use much more wireless network bandwidth than do voice calls or SMS, making the delivery of video more expensive. As more consumers sign up for video delivered over cellular networks (called unicast)—and watch it—operators foresee a time in the near future when their networks will inevitably become overloaded. The cost of delivering video on demand to individual user devices is simply too burdensome and expensive on a point-to-point network architecture. Thus, the implementation of systems that can broadcast the same information to virtually all users simultaneously provides the operators a way out of this dilemma.

There is a more complete examination of the contending standards and architectures for mobile video later in this chapter under video technology.

MUSIC

The number of music-enabled phones is increasing very rapidly. Most of the newer medium-to-high end phones are now music enabled. Although their quality might not match that of iPod, they do offer an alternate solution to users who want to carry just one personal portable device, a phone. The introduction of the Apple iPhone has established a new high-water mark for the handset industry to aim for with the integration of MPEG-3 (Moving Picture Experts Group 3)/iPod players into a unified cellular handset device. The value-added contributions pioneered by the iPhone include an innovative UI that simplifies and improves the navigation and control of all the devices' functions. As discussed above in the UI section, we believe the iPhone is just a good next step on a continuum of improvement in the user experience and certainly with the seamless integration of music and stored entertainment as a new domain of the handset. If we distill the iPhone to its core essence, it essentially establishes the cellular phone into a mass storage edge-caching device as well as being a communications platform.

Some of the most popular and best-selling devices are music phones. Over time, music sales on handsets will start to overtake that on the iPod and similar devices. There are already some indications of this trend. The first MP3 songs sold to phones were introduced in the summer of 2003 in South Korea. The International Federation of the Phonographic Industry (IFPI) said that, for 2004, the total sales of music to mobile phones worldwide was so trivial that it did not bother to break it down (i.e., much less than 1% of all digital music sold). However, on the back of enormous growth, for 2005, the IFPI reported that MP3 full-track music sold to mobile phones was "40% of all digital music sold." By the first quarter of 2006, the IFPI reported that "already half of all music sold goes directly to mobile phones." Mobile music is a fast-growing revenue stream for record companies, accounting for almost half of the US$ 1.1 billion revenues made from digital music in 2005. Japanese operator KDDI launched its full-track download service in 2004 and crossed the 30 million sales mark in December 2005. Over 110,000 songs are available for downloading. Vodafone included Universal's catalog into its library of full-track downloads raising its portfolio to 600,000 tracks available in 21 territories. In the United States, Sprint launched the first full-track dual download service, offering a catalog of 250,000 songs. The Sprint service is based on Groove Mobile's music delivery technology and will rely on Sprint's newly introduced high speed 3G networks (Figs. 6.2 and 6.3).

AUDIO

As is the case for any data resource, multimedia content can be compressed by removing signal redundancies, gaps, and signal frequencies inaudible to human hearing, using specific coding techniques. Compression efficiency has dramatically evolved with the improvement of processing technologies and DSPs in handsets. Today, it is possible to compress an audio file, preserving its original quality, with an impressive compression rate of 8. This means that an audio file with 8 MB in real format could be coded to fit in only 1-MB format for transmission over bandwidth-constrained channels. In addition, as processing power for mass-market phones shift to ARM9+ handsets, voice recognition capability is possible on handsets, and thus speech-based (multimodal) applications like the ones from V-Enable and VoiceBox will become prevalent.

GPS/LOCATION-BASED SERVICES

Global positioning system satellite capability is starting to make significant im-pact on the wireless broadband market. These satellites are contributing

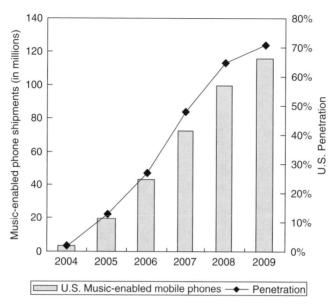

Figure 6.2. U.S. music phone penetration. *Source*: Lehman Brothers.

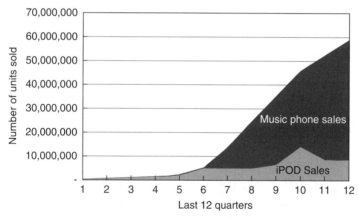

Figure 6.3. iPOD versus music phone sales—last three years. *Source*: SMLXL.

network clock-timing signals to base stations to coordinate network timing among multiple sites. The GPS has also been leveraged to provide emergency location information for the E911 notification extension into the cellular domain and is now extending its powerful contributions into the realm of location based services (LBS) for both commercial and consumer applications. Sprint is the first operator to open up its LBS APIs to developers on a selective basis. Verizon is also working to become more developer friendly for the creation of LBS applications.

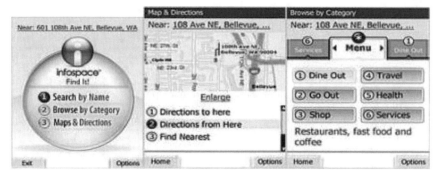

Figure 6.4. LBS-powered application from Infospace running on Sprint's Network. *Source*: Infospace.

Because both AT&T and T-Mobile are GSM-based networks, they presently rely on terrestrial radio-ranging triangulation technology to resolve position location for both E911 and LBS. As such they are disadvantaged in the integration of GPS-based services into their core networks. As they implement 3G-based WCDMA network upgrades, GPS becomes integrated into the fabric of their core network, thus reducing this disadvantage. In the meantime, it will be challenging for developers to build LBS applications and services for these operators (Fig. 6.4).

It should be noted that Nextel has been running commercial LBS for the enterprise and federal sector for many years. Besides local search and enterprise applications, LBS is also becoming part of the social networking phenomenon. Recently, Loopt, a start-up focusing on LBS-based social networking applications was launched on Boost Mobile. Several other mobile social networking companies are also looking to make LBS a common feature (however, they require blessing from the carrier to move any further) (Fig. 6.5).

Codeless GPS/Spectral Compression Positioning

In addition to traditional CDMA network–based assisted GPS systems, there are emerging technologies that hold promise of highly accurate universal location data for people, products, and vehicles, regardless of outdoors or indoors. Loctronix is developing next-generation technology for seamless, universal tracking and navigation, called Spectral Compression Positioning™ (SCP). It uses techniques based on "codeless GPS," but has far more advanced signal structures. SCP captures both standard and military GPS signals and then processes them on the same simple circuit that receives signals from autoconfiguring differential beacons. The system is highly resistant to multipath, uses little power, and can run either in hardware or as software-defined radio. Plans include deploying this system as an underlay 40 dB below WiMAX and 4G carrier systems as well as in

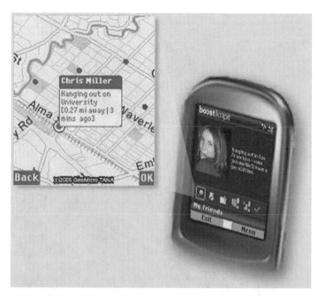

Figure 6.5.   Social networking based on LBS. *Source*: Lehman Brothers.

industrial and consumer products. Figure 6.6 illustrates the challenges facing traditional GPS-based positioning versus the SCP approach.

The following figure depicts the differences among traditional GPS, real-time locating system (RTLS), and the Loctronix SCP architecture (Fig. 6.7).

The use of SCP holds the promise of improved performance and substantial cost reductions for the applications and equipment required for LBS and tracking of inventory and resources across a wide range of industries. Field trials are presently under way with major OEM (original equipment manufacturer) electronics manufacturers in Japan at this writing. Additional information can be obtained by the reader at www.loctronix.com (Fig. 6.8).

MESSAGING AND HOSTING

The advent of ubiquitous broadband wireless networks will drive in an era where many of the narrowband services and messaging techniques that were developed to compensate for limited bandwidth and expensive transport will be replaced by universal, low cost e-mail, fully featured, and indeed even augmented by rich media, including image, sound, and video. It is likely that this major change in access bandwidth at the edge will be the death knell of the "Blackberry" unless they elect to migrate their technology to purify an architecture that supports pure IP access and transport. The following list comprises the key elements of the

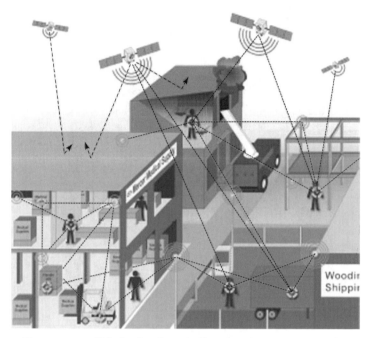

Figure 6.6.   GPS shadowing challenges. *Source*: Loctronix.

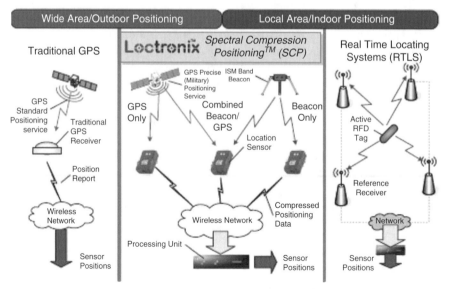

Figure 6.7.   SCP-enhanced GPS. *Source*: Loctronix.

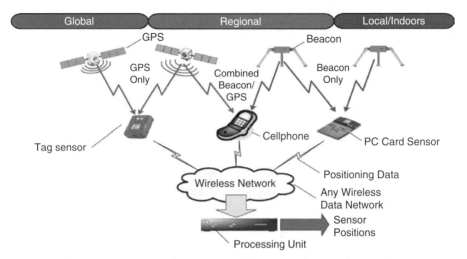

Figure 6.8.   Loctronix system architecture. *Source*: Loctronix.

messaging industry that will expand their feature sets and extend their utility value under a true broadband network environment.

- E-mail
- Unified messaging
- Instant Messaging (IM)
- Multimedia messaging
- Archiving e-mail uploads and downloads
  - Online real-time accessible storage
  - Storage area networks (SAN)
  - Yahoo and Google have already implemented this service
- Software as a service (SaaS)
  - Formerly what was called the application service provider (ASP) model that collapsed in the Internet bubble bursting is coming back as SaaS as universal broadband access has become a reality.
  - Enhanced to include online outsourced resources for a wide range of professional support services in addition to user applications.

INTERNET 2/WEB 2.0 SOCIAL NETWORKING

The wireless broadband future will also be a significant driver for increasing the utility value of a wide range of social networking businesses and organizations. Social networking has already been proven to cover not just the personal lives of the FaceBook generation but also to empower professional connections through the growing web of relationships that are daily expanding over LinkedIn and similar social and professional networking sites.

User-Generated Content

In the midst of the rapidly expanding "blogosphere"* the blog search engine Technorati identified more than 106 million blogs, circa September 2007, and the Time Magazine selected all of us, "you," as its person of the year in 2006 to acknowledge this extraordinary expansion of the Web's value as a tool of inter-action between and among all persons, regardless of personal acquaintance or introduction.[†]

The self-publishing of a wide variety of content using the Web as the ultimate "vanity publishing" mechanism has led to the widespread dissemination of opinion, ranging from the thoughtful and profound to the slanderous and the profane. All this new or pilfered information launched into cyberspace by millions of citizens seeking to reach an audience for their respective purposes, be they for commercial, political, philosophical, psychological, humanitarian, predation, personal, or cor-porate agendas to name a few.

Sorting through this enormous stack of "stuff" has been made almost prac-tical by the expansive growth of the search industry. Led by Google and followed by numerous generic or specialized search engines, the Internet does tend to give up its treasures to those with the patience and persistence to seek them out.

USER DEVICE FORM FACTORS

In the next three to four years, mass-market cell phones will fundamentally remain the same in size though screen size and resolution will keep on improving. Hybrid devices will emerge that will cater to certain niche segments, but as a mass-market device, cell phones will likely retain their current form factors during this period. What will change are slightly bigger screens or designs that have virtual keyboards or projection screens from the device or tighter integration of voice into the handsets for multimodal communication. As the base processor moves from less than ARM7 to higher than ARM9 over the course of the coming three to four years, one could have smartphone-like functionality in average handsets. Another area that might see some progress are expandable (folded or rolled) displays, which will provide a bigger display capability, coupled with projected virtual keyboards that will largely mimic laptop personal computers.

Over the next decade as on-demand flexible manufacturing techniques become widely available, it is likely that user terminal form factors will evolve dramatically.

---

*The term blogosphere was coined on September 10, 1999, by Brad L. Graham, as a joke. It was recoined in 2002 by William Quick, Wikipedia.org/wiki/Blog.

[†] See Wikipedia.org/wiki/Blog.

PROCESSING POWER

Chip manufacturers are looking to more efficient architecture designs, including power management systems. ARM, for example, has recently introduced Intelligent Energy Manager (IEM) technology, which optimally balances processor workload and energy consumption while maximizing system responsiveness for better end-user experience. In addition, the majority of CPU vendors—including Texas Instruments, Intel, and Motorola—are now integrating power management systems to reduce the energy consumption of their CPUs. For example, in Intel's recent handheld CPUs, power consumption is reduced using idle/sleep modes and turbo mode that provides the flexibility of changing clock speed. For example, a device using a 400-MHz CPU can conserve power by running 200 MHz for downloading a video file but can jump up to 400 MHz to view the video. In a similar fashion, wireless signal processing using DSP can be reduced to idle mode when the modem is only tracking the transmission signal and to the maximum when a broadband communication is established building a one-chip solution that combines transmission signal, code and data processing, and memory management on a single die. This dramatically reduces power consumption by decreasing the number of unwanted interfaces and discrete components. In future-generation processors, manufacturers will focus more on reducing energy consumption rather than racing for higher speed.

DATA MANAGEMENT

Increases in the performance and capacity of low cost data storage devices will enable several new features and applications. For example, operators will be able to download their entire content catalog onto the handset by downloading the content in the background while the handset is idle, so that when user is ready to interact with the content, it is already on the handset. All that is required to access specific content is to send a request to the server, enabling digital transfer rights (DTR) payments and the provision of a public key to facilitate the specific transaction. Functions such as preview and unlocking of downloaded files will happen right on the device without much server interaction or network burden. Similarly, videos or other multimedia content can be stored on the device for later processing or uploading to the network. Additional storage also means capabilities such as music- and video-enhanced applications and interactive gaming can finally be brought to the market. In addition, increased storage means one can take higher resolution pictures from the handset. Storage can also be used for storing other documents such as in Word or PowerPoint. With 3 to 10-GB+ drives, one can carry a whole lot of documents and content with them.

CAMERA/VIDEO MANAGEMENT

Today, an embedded camera has become a common feature of many handsets worldwide. First-generation camera phones employed complementary metal oxide semiconductor (CMOS) image sensors in their camera modules. Handset manufacturers initially placed a high priority on low power consumption and low cost modules at the expense of image quality that did not exceed cells in frames (CIF) quality image (101,376 pixels). This choice was justified because, at the time, wireless networks could not efficiently transmit bandwidth-hungry quality images and screen technologies were not developed enough to view such images. Given these conditions, CMOS was the technology of choice over opposing charge-coupled device (CCD) technology.

In the mobile handset market, CMOS sensors have many advantages over CCD technology, including their cost, power, and size. Nevertheless, the most noticeable advantage is the ease of integration, enabling CMOS sensors to combine image sensing, digital logic, and memory functions onto a single chip, whereas CCD modules require extra chips to process and convert their analog signals into digital.

Although camera phones are currently gaining widespread popularity within mobile handset market, the low quality of the images produced by these devices is slowing down their use as a booster of wireless services such as multimedia messaging or picture messaging. It is said that only a handful of consumers (less than 3%) share pictures taken on their mobile phones. Furthermore, more than 50% of camera phone users used the camera feature only very occasionally or never at all. Consumers have come to expect a certain level of image quality and are not prepared to pay for wireless transmission, sharing, printing, or even storing photos if picture quality does not match their expectations. The initial reaction of the industry has been to increase the number of pixels in camera phones. The technology is now enabling the creation of high performance sensors supporting megapixel and higher rates with decent volumes and reasonable pricing. However, the increase of pixel rate on its own does not translate into better image quality. Indeed, the number of pixels only has an influence on the level of detail carried by the picture, which is useful for printing large-format pictures, wide-screen viewing, and digital zoom-in functionalities. The increased number of pixels does not improve other important parameters such as color richness, contrast, or sharpness, but results in larger images, which increases the need for memory and digital processing capacity as well as bandwidth for wireless transmission.

Many factors influence the quality pictures taken by a camera phone, these include lens, color filters, the size of the image sensor, supporting DSP, and software for image processing. With the advent of 3G, storage capability will increase and carriers will allow for higher resolution pictures to be stored and transmitted.

MOBILE ADVERTISING

Looming in the wings of a number of strategic planning offices in the wireless industry is the issue of advertising into the handset universe. The potential to reach 3 billion outlets for advertising impressions is too significant an opportunity to go untapped. As cellular operators continue to face erosions in their traditional voice ARPU, they will be highly motivated to entertain various forms of advertising schemes onto their platforms.

There is another camp that is seeking to leverage advertising as the means of delivering free services in exchange for engaging with selected and targeted advertising. MetroFi in the municipal WiFi space has already been relying on this model, but it is still too soon to judge if it is able to earn sufficient returns, given the user community is still small and cannot deliver cost-effective impressions across millions of users. Cellular on the other hand suffers no such infirmity of potential impressions.

Coupled with the recent advances in network-based sniffers, which are able to pair users with advertising content that is matched to their demonstrated preferences, the potential for advertising in the mobile wireless market is enormous. The ability to tie user's personal preferences to LBS and following their specific requests to either opt-in or opt-out of pushed, contextually prescreened advertisements will create one of the most powerful advertising media ever created.

Companies such as Feeva Technology Inc. are building the engines that will both maintain the confidentiality and anonymity of Internet users, while also establishing preference profiles that can be used by targeted advertising agencies seeking to implement the vision of the extremely high value-added nature of personally targeted advertising described above. We anticipate that there will be an explosion of both intrusive and "more individual friendly" tools to leverage relationships between wireless subscribers, Internet users, and the future broadband wireless ecosystem that is looming ever closer to affecting our daily lives. As the old saying goes, "You can run, but you cannot hide." The twenty-first century versions of privacy are going to take us into unchartered waters, and it is incumbent on all of us as citizens and consumers to pay attention to the public policy debates that will inevitably rage over these matters, but against which our only participation is likely to be what is paradoxically marshaled over the blogosphere wherein we are all inexorably linked to a common fate; no secrets!*

VOICE

Voice is dead, long live voice: VoIP, Skype, SIP, VIM, etc.

- VoIP (telephony quality)
- VoIP (high fidelity)
- Voice conferencing
- Voice messaging
- Voice IM
- Voice to text
- Text to voice
- Audio coder-decoders (CODECs)

---

* For a more detailed treatment of this topic, please refer to "Mobile Advertising: Supercharge Your Brand in the Exploding Wireless Market" by Chetan Sharma, Joe Herzog, and Victor Melfi (John Wiley & Sons, 2008).

Voice over Internet Protocol

One of the most important developments that complements the rapid expansion of broadband IP is the ever-refining capability of VoIP and the ability to efficiently replace traditional circuit-switched voice telephone services. VoIP began its life as a technique used by computer enthusiasts to connect to each other via PC-to-PC over the public Internet. It was quickly adopted by the competitive long-distance industry as a low cost alternative to traditional tandem switch–interconnected long-distance circuits, especially over international routes. Over the ensuing years since the collapse of the competitive telecom sector in 2000, VoIP developers have continued to refine the technology, and coupled with the growing availability of broadband services over DSL and cable modem, it has been gaining rapidly as a stand-alone service provided under the "bring your own broadband" business model that has proliferated throughout the industry.

A large number of VoIP service providers have sprung up across the country, some small "mom and pop" operations are doing very well, while the largest specialized VoIP service providers are struggling in the wake of the Comcast Digital Voice initiative and the similar offerings of the other cable system operators. The challenged include Vonage, which is fighting to maintain its market position in the face of negative legal decisions asserting that it has infringed ILEC-owned intellectual property. Recently, SunRocket, the second largest stand-alone VoIP service provider, closed its doors on short notice, stranding almost 259,000 subscribers without telephone service. There are literally hundreds of VoIP service providers addressing both consumer and enterprise VoIP services throughout the world, serving up a large variance of service quality and prices.

The widespread adoption of Skype VoIP services throughout the world is a solid example of how the legacy telephone network is essentially unable to keep pace with the advances in technology, which are far ahead of the regulatory framework that is completely inadequate to oversee the widespread effects and implications of the Internet coupled with VoIP technology.

The combination of broadband IP voice services seamlessly connected to the legacy telephone networks, which allows for a call to be originated in one domain and terminated in the other, is a hugely disruptive capability. Given that PC-to-PC connections are flat rate and free of metered service charges, regardless of their location anywhere in the world, and that termination to traditional telephone numbers only costs a few cents per minute under these new hybrid service providers, the impact on traditional telephone company economics is devastating.

Voice generates approximately 85% of all revenues in the telephone industry. When this figure is collapsed to reflect the decrease in revenue from the legacy tariffs of the traditional telephone industry, treating voice as data, the top line of the entire industry faces a potential reduction of almost 80% of its revenue contribution. The results of this trend are destined to be extremely disruptive to organizations that are wedded to large top line income numbers. We predict that the voice elements of the telephone business will continue to be profitable;

however, it will be on a much smaller gross revenue base and will inevitably reflect the operations of smaller, leaner, and more flexible organizations.

Cellular/VoIP Hybrids

There are some main themes emerging in the cellular industry about the pros and cons of deploying a WiFi data or VoWLAN solution integrated with cellular handsets. Below are listed some pros and cons of this hybridized technology development, with breakouts for both the consumer market and the business market, since these have vastly different requirements.

Enterprise—pro

- Unified billing
- Better in-building coverage
- VoWLAN
- Helps customers migrate from LAN data usage to cellular data usage—
- Local wired telecom displacement
- Cost savings of providing voice service
- Reduce churn

Enterprise—con

- Customer service hassles of not having total control of the network
- Wireless service providers don't want to be in the PBX business
- WiFi handsets can be costly

Consumer—pro

- Wired provider displacement
- Customer loyalty

Consumer—con

- Added handset cost
- Relies on customer-supplied hardware with potential for high support costs

In 2007, an announcement by operator 3 in the United Kingdom is probably indicative of future cellular industry trends to come in the area of all-you-can-eat data plans. So far, only Willcom in Japan has really had a flat-rate plan, which it pioneered on its personal handyphone system (PHS) digital network. Operator 3, which was previously very much a "closed garden" operator, decided to open

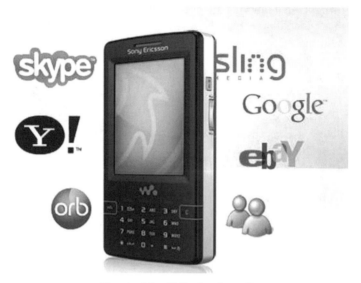

Figure 6.9. X-Series from 3.

things up at a fixed price. It is partnering with Skype, Slingmedia, Yahoo, Ebay, Microsoft, and Google to offer a full complement of applications (Fig. 6.9).

This approach, pioneered by 3, is likely to force the bigger players like Vodafone, O2, and Telefonica to rethink their pricing strategy in the coming months. The authors believe that there will be a spillover effect into the United States where Helio is offering a similar approach, but it is a very small player. However, it will take time before the U.S. "walled garden" operators truly open up their access networks, as they still face substantive competition. The Mobile WiMAX initiatives may still prove to be of sufficient threat to stimulate the Cellcos into action, but it is not likely to happen quickly.

VIDEO

The evolution of the "three screen" marketplace for video triggers what we termed earlier the "Negroponte reflux." Much of the video content that migrated into cable networks is now sneaking back out into the wireless airways, although in most cases at frequencies and with protocols that were not previously (except for 700 MHz) used for video distribution. Ironically, it will be in the 700-MHz bands that were formerly the upper domain of the UHF television broadcasters that are being placed into service for the Qualcomm MediaFLO video to mobile handset business, and the remaining portions of the band that will be auctioned by the FCC in early 2008 for the inevitable implementation of mobile broadband IP networks that will certainly be implemented with the intention of being fully capable of supporting both broadcast and unicast video content. AT&T

purchased the 700-MHz licenses consisting of 12 MHz of bandwidth that had been acquired by Aloha Networks for $2.5 billion in October 2007. Aloha Networks had been using the spectrum in DVB-H trials of video to the handset. AT&T has made no public pronouncements about its intentions for these newly acquired bands to the date of this writing, and is expected to seek additional 700-MHz spectrum in the January 2007 auction.

For several years we have been observing the march forward of IPTV from small-screen windows on our personal computers acquired using software applications from Windows Media Player, Real Networks, Adobe Macromedia, and Apple QuickTime. We have observed and enjoyed the migration of the Windows operating system onto our handheld personal digital assistants (PDA) and related devices, all steps on the trail to full-screen scale, nonpixilated, approximately 30 frames per second video content available on virtually any of our three screens, originating in either native digital video broadcasting (DVB) format or encapsulated in an IP stream as IPTV, or soon, with higher resolution as television over IP (ToIP). The ability to bring high resolution and full-motion video to all of the three primary screens that intersect with the user community on a consistent basis, will be essential to refine the service to acceptable levels that will drive demand into every corner of society.

VIDEO COMPRESSION TECHNOLOGIES

Video CODECs

There are a number of video compression CODECS that have been applied to the digital conversion of video over the past 20 years. The leading contenders for likely intersection with the broadband wireless networks of the near future will include the following technical standards:

- MPEG-4 (in all of its many versions)
- H.264/AVC (advanced video coding) (the ITU standard evolved from H.263)

MPEG-4

MPEG-4 has a large assortment of features and standardized interfaces that are still evolving at the present time. The technology behind MPEG-4 is proprietary, and the rights holders have pooled their interests into a licensing authority called the "MPEG Licensing Authority, LLC" (MPEG LA), which is a firm with the responsibility to seek licenses from those who use its technology. The range of standards under its administration includes the following:

- MPEG-2
- MPEG-4

- IEEE 1394
- DVB-T
- AVC/H.264 standards

MPEG LA is also the likely licensing authority for the emerging VC-1, ATSC, DVB-H, and Blu-Ray technologies. The MPEG LA organization is headquartered in Denver, Colorado.

Advanced Video Coding/H.264

The AVC/H.264 standard delivers twice the compression power of the earlier H.263 standard. The standard is also referred to as MPEG-4 Part 10. The ITU Television Video Coding Experts Group has responsibility for its development in concert with the International Organization for Standardization (ISO)/International Electrotechnical Commission (IEC) MPEG. It is the product of these two international groups, and as such is also referred to as the product of the Joint Video Team (JVT). Work on the standard was completed by the JVT in May 2003.

Wavelet Compression

Wavelet compression is presently not incorporated into any of the existing standards, but it holds great potential for future highly compressed video and image compression. In addition to supporting extremely low bit rate, high quality images, the technology also allows for the same digital video stream to scale dynamically for multiple resolution screen environments, including full-size televisions, personal computers, and handheld devices. Wavelet compression has been demonstrated to deliver compression advantages of greater than 100:1 over current Joint Picture Experts Group (JPEG) and MPEG approaches to image compression.

There is a never-ending race to balance scarce network resources against an ever-growing demand for bandwidth to deliver large-size data files. The broadband wireless networks of the future will benefit from both increased capacity as well as more efficient data compression technologies. Below are two examples of wavelet compression techniques applied to the same fixed image comparing them against a standard bitmap and a GIF file (Fig. 6.10).

At the present time, there is a contest under way to determine which standard for handheld mobile video devices will predominate in the various global regions. There are presently four major contenders, each with a unique market position to leverage in pursuit of widespread adoption and potentially spread beyond their originating regions. Following is a description and overview of each of the leading standards, and a comparison between their technical specifications and deployment characteristics. One of the challenges facing this emerging market segment is the limitation inherent in the digital video streams that are optimized for a specific display device. As the market matures and full convergence evolves in earnest

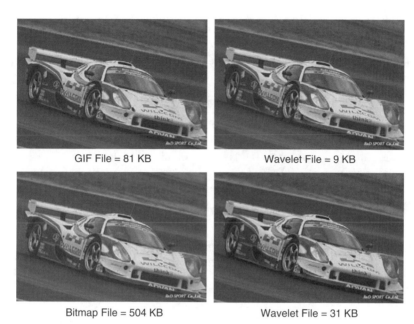

GIF File = 81 KB   Wavelet File = 9 KB

Bitmap File = 504 KB   Wavelet File = 31 KB

Figure 6.10. Comparison of wavelet compression images.

among the various service providers, the need arises for the same transmission streams to be dynamically scalable for display on any of the three digital screens that are defining the new video landscape: television, PC, and handset.

Listed below are the four primary global candidates for mobile video/IPTV.

- DVB-H—(Europe/ETSI) "digital video broadcast–handheld"
- DVB-SH—"digital video broadcast–satellite handheld"
- Qualcomm MediaFLO—(United States/proprietary) "Media Forward Link Only"
- CMMB—"China Multimedia Mobile Broadcasting", the Chinese mobile broadcasting standard mandated by SARFT.

DVB-H: (Europe/ETSI) "Digital Video Broadcast—Handheld"

The European Union has been a world leader in the development of next-generation standards for mobile digital audio and video services. The early adoption of what was known as the Eureka-147 standard for digital audio broadcasting (DAB) spawned a number of initiatives to deliver video to mobile vehicles and ultimately to handheld devices. The result of these efforts was the development of the DVB-H standard. It has been accepted as the best way to capture early market share in the mobile video market by many cellular operators in Europe and by early Verizon mobile video services in North America.

DVB-SH: "Digital Video Broadcast—Satellite Handheld"

This is the satellite-augmented hybrid version of DVB-H. It utilizes geostationary satellite transmitters to provide wide area coverage and terrestrial gap fillers to provide indoor and ubiquitous coverage in high density urban areas with frequent line of sight blockages of the satellite. Nokia and Alcatel Lucent have been promoting this architecture, which supports digital video transmission direct to small handheld devices and which also uses terrestrial amplifiers to support in-building penetration of signals in urban areas. The satellite feed provides universal coverage over the service area without the need for terrestrial cell sites in all service locations. Clearwire has announced trials using the DVB-SH technology with spectrum holdings at 2.0 GHz of its sister company ICO Global Communications, which can also be used terrestrially under its Ancillary Terrestrial Component (ATC) authorization.

Qualcomm MediaFLO: "Media Forward Link Only"

Qualcomm has developed a proprietary solution for mobile digital video both as a product and as a service. Qualcomm acquired nationwide spectrum in the early 700-MHz auction to provide a video broadcasting service that is made available under a resale approach to cellular carriers in the United States. In markets outside the United States, Qualcomm is seeking to proliferate its solution in other frequencies and to openly license manufacturers to support its standard. MediaFLO has captured the leading position among the mobile video broadcasting with the existing cellular operators in the United States, including service agreements with both Verizon and AT&T. Alltel and T-Mobile USA have both announced trials.

StiMi (China/SARFT): "Satellite and Terrestrial Interactive Multimedia" now Named CMMB

The Chinese government has determined that mobile video services are to be a national priority. The State Administration for Radio, Film and Television (SARFT) has developed a domestic standard called StiMi, which is a satellite-based service operating over three 8-MHz channels in a 25-MHz bandwidth at 2600 MHz in the S-band. The system also supports additional spectrum feeds within the UHF bands that in addition to the regeneration of the satellite signals can also be fed from the terrestrial gap fillers, which are integrated into the system. The hybrid satellite terrestrial network enables both ubiquitous national coverage via satellite and indoor reception by handheld devices from the terrestrial gap filler transmitters.

The system is designed to provide 20 video streams at 256 Kbps and 20 audio streams at 128 Kbps within each 8 MHz of available bandwidth. SARFT has teamed with the Hong Kong–based subsidiary of Echostar for the design and deployment of the S-band satellite that will deliver the nationwide service under an outsourcing agreement to SARFT for delivering signals both directly to user handsets and to the terrestrial gap fillers. It is the stated objective of SARFT to have the system operational in advance of the 2008 Olympics in Beijing. The network is called China Mobile Multimedia Broadcasting (CMMB).

The technology has been developed under contract with the Chinese SARFT by Terrestrial Interactive Multimedia Infrastructure (TiMi technology) of Beijing, which was founded by Dr. Hui Liu, who was introduced to the readers as the pioneering developer of the OFDMA technology at Broadstorm and Adaptix. Earlier in his career at C-Will, he was instrumental in the original development of TD-SCDMA, which has become the domestic 3G standard of China and which along with Mobile WiMAX that has derived directly from his earlier work has been recognized by the ITU as one of the (only 6) internationally approved standards for mobile communications under the IMT-2000 specifications. This is a truly remarkable achievement by a true visionary in our industry, and we predict one of the future legends of our industry. It is remarkable he has achieved all of this before the age of forty, so we can also anticipate even more contributions from this remarkable man.

FIXED DIGITAL VIDEO SERVICES

IPTV/Digital Video/ToIP

- Internet-based IPTV
  - On-demand video downloads
  - Video messaging
  - Video conferencing
  - Video surveillance
  - Video blogs/You Tube
  - Self-produced video
- Commercial terrestrial broadcasting digital conversion
  - DTV
  - HDTV
- Satellite broadcasting
  - To head-end facilities (ISP, Cable, Cellco)
  - To individual users (DBS/DTH)
- Corporate video
- Distance learning
- Telemedicine
- HDTV movie and event distribution to theatres

Internet-Based IPTV Services. In addition to the ability to tap into streaming video content on the Web, the most significant market opportunity facing the content providers who can take advantage of the growing availability of true broadband capacities throughout the world is the ability to time-shift access to content under schedules that users define. The Internet is still primarily a point-to-point network, thus the ability to simulcast shared IP content to multiple addresses is limited by the number of ports available for users to access simultaneously to participate in real-time content delivery. One of the major benefits that broadband IP wireless networks will enjoy is the inherent ability to deliver a data cloud over the entire footprint of the wireless system with simulcast broadcasting of content to all users who wish to have access.

ON-DEMAND VIDEO DOWNLOADS. Ultimately, content providers will be able to deliver hosted content on demand to any device that has sufficient bandwidth to support the reception of video. Essentially, the future for on-demand video will leverage the global Internet as a massive "TiVo* type" platform. The on-demand downloading of content will also be subject to many forms of digital content rights purchase or rental agreements. Outright purchase for personal use, onetime playback rentals, and all forms of various commercial rights to use that can be safely administered by Digital Rights Management (DRM) systems are integrated into service provider networks.

VIDEO MESSAGING. The mass-market adoption of digital cameras and personal digital video cameras and their widespread proliferation into cellular phones for both still and moving image will result in a dramatic expansion of video e-mail. Friends and families will no longer be limited to sending only still photos over the Web, but will be able to provide real-time video messaging regardless of their location. Broadband wireless networks will redefine the balance of upstream and downstream capacity, given the rapidly increasing ability to consume substantial bandwidth by users at the edge of the network.

VIDEO CONFERENCING. Personal and corporate video conferencing will achieve new levels of convenience, cost-effectiveness, and quality with the introduction of widely available true broadband services. Full-screen broadcast quality video will quickly make obsolete legacy "window pane" video conferencing solutions and also eclipse the Web-based solutions that are constrained by the number of simultaneous users and display in small windows on computer screens.

The ability to engage face-to-face, in a high resolution image environment, between every participant on a videoconference will revolutionize the utility value of these services. The personal travel habits of the modern business practitioner may come under severe competition using these enhanced services. In a world facing dramatic climate change based on proliferation of carbon emissions into the atmosphere, the pressure to offset unnecessary travel will become immense.

The use of shared electronic whiteboards, working papers, and the ability to use real-time video to examine specific details of products or training materials will also support the widespread proliferation of enhanced video conferencing. As the cost of broadband access and transport reduces to anticipated commodity levels over the coming decade, a renewed interest and improvement to this basic concept

---

*TiVo is the brand name for the pioneering digital video recorder from Teleworld, Inc.

is expected, which along with decent quality personal videophones has been an "evergreen" promise within our industry.

VIDEO SURVEILLANCE. Government and commercial applications for video surveillance is increasing rapidly. The market for IP-based closed circuit television security monitors is being driven by a need to increase the resolution of the cameras, and to integrate facial recognition software to attempt to intercept security risks in advance of their perpetrating any illegal acts. Municipal governments have been constrained in terms of how many cameras they could deploy in advance of the availability of broadband network facilities. The concept of public–private partnerships for municipal wireless systems has been proliferating nationwide, typically for WiFi mesh implementations.

These Muni WiFi networks and related public safety versions in the dedicated public safety bands at 4.9 GHz have been finding success in the smaller and midsized cities. Recently, the Muni WiFi initiatives in several of the major cities have fallen upon difficult times, and pioneering companies in the space such as EarthLink and MetroFi are pulling back from earlier plans as they have become exposed to the reality that public use and Internet subscriptions over these networks are not meeting original expectations. Regardless of whether the future of public space video surveillance is carried over public-shared WiFi or WiMAX networks or via dedicated public safety infrastructure, the rapid expansion of video surveillance is inevitable.

VIDEO BLOGS/YOUTUBE/SELF-PRODUCED VIDEO (WEB 2.0 CONTENT). The success of YouTube and the proliferation of self-produced video content in support of personal Web sites, Really Simple Syndication (RSS) video blogs, and the increasing use of video to support Intranet applications for business, training, and customer service applications will be enhanced and driven by the expansion of broadband IP wireless networks of all types.

Flickr, a Yahoo company, is typical of a growing number of Web portals providing photo and image storage and manipulation online. It provides the ability to upload from computers, cameras, mobile phone cameras, and via e-mail. Services also include tools for the organization of image collections with date, time, and location tags. There are also controls to enable the sharing of photos with authorized users and a number of products to enable the creation of cards, postage stamps, and prints. Other examples of photo (and some video) storage and sharing sites include the following:

http://www.shutterfly.com/
http://photobucket.com/
http://www.snapfish.com/
http://www.myphotoalbum.com/
http://www.smugmug.com/
http://www.fotki.com/us/

This has become a crowded field, and a number of sites have either ceased operations or announced that they are taking no new subscribers and plan to discontinue operations in the near future. The sector has evolved three distinct business models that are either advertising supported or subscription based or adjunct services to other businesses.

Among the video sharing and storage Web portals, YouTube, now owned by Google, is undoubtedly the market leader, but it is joined by a large field of other video-centric services that provide social networking of video images. The authors have identified over 103 separate video storage and sharing sites at the time of this writing. Listed below are some of the leading sites.

http://dave.tv/
http://www.esnips.com
http://crackle.com/
http://www.jumpcut.com/
http://home.myspace.com/
http://tv.oneworld.net/
http://www.ourmedia.org/
http://www.scenemaker.net/
http://www.twango.com/
http://uncutvideo.aol.com/Main.do
http://www.vidmax.com/

Image

The integration of digital cameras into cell phones and future broadband IP user devices has ensured that images are forever wedded to the wireless industry. The convenience and immediacy of being able to snap a photo of anything of interest virtually instantly has resulted in capturing some of the more memorable glimpses of natural and artificial tragedies, disasters, and news events, and has thus revolutionized electronic news gathering and provided new witness to world events. From the profound to the trivial, we are snapping away. The ability to send these images to our respective storage locations, typically online, is another driver of the need for increased data throughput and system-wide capacity, as we challenge our networks with bit-rate intensive services far beyond what planners anticipated for delivering primarily voice services.

Examples of typical image file transfers that will increase future demand for real-time bandwidth and the administration of our image archives include the following:

- Photo storage and distribution
- Image search and recovery
- Commercial image DRM registration and purchase

- Maps
- Field workers (pothole reports, infrastructure logging, site surveys, etc.)
- Technical manuals for field access by technicians, service staff, and engineers

TRADITIONAL DATA SERVICES

The wireless industry had historically supported a wide range of commercial applications for radio communications. As we move into an era of large-scale, shared IP-centric infrastructure, there will be a number of overlaps and indeed repurposing of what will rapidly become obsolete applications-specific radio networks. The creation of secure VPNs with firewall and access protection delivered under service level agreements (SLAs) defines strict QoS requirements for the service provider to deliver under contract to the user, whether a large corporation or a single individual.

We anticipate the migration of many purpose-built, user-specific radio networks to either go dark or be repurposed to some higher use via a spectrum refarming process in the future. Among the categories of dedicated RF and data networks are the following:

- SCADA
- Public safety
- Banking/ATM/IEMD financial applications
- Contextual database services
- IP-SS7 (in-band IP signaling for VoIP to telephony networks)

SCADA (SUPERVISION, CONTROL, AND DATA ACQUISITION)

SCADA networks are widely disbursed across the world to provide mission critical information for a wide range of applications, including resource extraction (well and pipeline monitoring), natural resource management (dams, waterways, water distribution systems, sewers, etc.), security monitoring (alarms, report by exception, polling, camera, etc.), and sensors (heat, cold, intrusion, fluid or pressure levels, light switches, etc.).

All of these applications require both purchase and maintenance of fairly low volume and expensive equipment that is only required to transport what are now considered extremely narrowband data messages, typically between 4.8 and 19.2 Kbps. The arrival of ubiquitous broadband shared IP networks will revolutionize both the efficiency and the utility value of SCADA over broadband, improving reliability (commercial networks are monitored on a full-time basis $7 \times 24 \times 365$ and have strict restoration time if an outage is detected); such oversight and service restoral response is rare among private SCADA networks, and if it does exist, it is likely to prove cost-prohibitive as we move forward into a period where higher performance, most cost-effective and reliable alternative is present.

GAMING

The gaming community worldwide is a major consumer of broadband services and is one of the segments of the application drivers for increased wireless broadband availability that simply has no limit to its appetite for bandwidth. As more bandwidth is available, the realism of the gaming graphics will continue to be refined into what are ultimately anticipated to be full immersion, virtual reality experiences, which will require hundreds of megabits per second to deliver in real time.

We have progressed from Pong on our 8088 monochrome DOS machines with ASCII graphics to a level of realism (or fantasy environments) that is simply stunning to behold in its details. The screen sprints below demonstrate the case far better than words alone can convey (Figs. 6.11 and 6.12).

SENSOR NETWORKS

The extension of sensor monitors and network connections to enable safe and reliable monitoring of loved ones in elder care or children in day care has become a significant area of interest for the general public. Parents and children at both ends of life present many challenges to their effective supervision by those responsible for their well-being. This field of application is another growing driver mitigating for the availability of broadband wireless networks. The ability to provide real-time monitoring of both the environment and the personal physical condition of those in our care will be a significant benefit to all parties, including the caregiver.

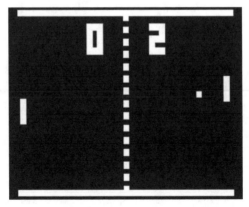

Figure 6.11. Pong, video games, circa 1972. *Source*: Pong.

Figure 6.12. Screen print of Play Station 3—Formula One Championship. *Source*: Formula-One-Championship-2007-PS3.

Personal health and safety monitoring has previously been limited largely to "panic button" transmitters, which required users to be able to press the button (unreliable if they are unconscious), and lack in information content to help the caregivers triage the situation promptly and marshal an appropriate response. With the pendency of broadband IP-based networks, a wide range of sensor and communications tools will become available and cost-effective enough for families to very likely embrace the comfort of knowing in real time the constant state of well-being of their loved ones, whether they are in their own homes or in a caregiver facility.

Following are several types of sensor applications that will become part of the wireless broadband ecosystem, as the industry matures.

- Personal/medical
  - Heart and health monitors
  - Elder care video and live audio supervision (stay in their own home longer)
  - Child care facilities supervision and personal monitors
- Environmental
  - Water rights monitoring
  - Air quality
  - Weather station monitors
  - Weather cams
  - Water quality

- Flood warning and Tsunami monitors
- Fish and game monitoring and tracking
• Intrusion and security detectors
  - Burglar and fire alarms and home video monitors
  - Bilge and security monitors for private vessels

7

THE EMERGING INFLUENCE OF THE COMPUTER INDUSTRY

WIRELESS LOCAL AREA NETWORKS GROW UP AND OUT: MUNICIPAL WiFi

It will be extremely interesting to see how effective the computing industry's encroachment into the telecommunications space will continue to be in providing the public with access to truly open broadband IP wireless networks. These new open networks will empower users to dynamically, session by session, select the underlying content or specific service provider on the basis of their personal preference, price, quality, and type of unique service desired. The success and widespread proliferation of WiFi (Wireless Fidelity*) hot spots, followed by the increasing momentum of municipal-scale WiFi network deployments, will provide a baseline against which open access and open network services will be measured by consumers against traditional closed network operators.

The mass manufacture of wireless local area network (WLAN) equipment that conforms to the 802.11(b/g) standard resulted in the development of very low-cost wireless data components and devices, which manufacturers integrated into a wide array of products, including the WiFi industry lift provided by Intel when it incorporated the technology into its laptop chip sets. Although the suppliers of discrete WiFi chip-level solutions initially suffered from Intel's move, the industry

---

* Wireless Fidelity (WiFi) is the brand name the WiFi Alliance, the trade association that promotes and controls the interoperability of WiFi devices, uses to identify compatible standards-conforming products.

Wireless Broadband. By Vern Fotheringham and Chetan Sharma
Copyright © 2008 the Institute of Electrical and Electronics Engineering, Inc.

has continued to grow exponentially, thus providing a wide range of new applications and products that require discrete WiFi system-on-a-chip solutions.

The expanding WiFi market is driving a continuous process of refinement and improvement to the original standard. Increased speed was captured under the 802.11(g) enhancement, which increased the basic operating speed of the system from 11 to 54 Mbps. In addition, the adoption of 802.11(a) extended the standard to include the European HyperLAN standard and the UNII (Unlicensed National Information Infrastructure)* frequency bands. At the time of this writing, the latest enhancement to the standard, 802.11(n), is in the final stages of review by the IEEE working group. This latest enhancement will increase the baseline data rate to 108 Mbps and potentially up to a theoretical maximum throughput of 300 Mbps in a perfect environment, using all the optional MIMO (multiple input/multiple output) antenna features incorporated in the standard. Perhaps of greater utility significance is the inclusion of quality of service capabilities into the WiFi standard, which enhances the limitations of the "best efforts" approach of the original underlying Ethernet protocol conforming to 802.11(e).

In addition to the tremendous growth of the WLAN market for both consumer and enterprise applications, numerous manufacturers have repurposed WiFi technology into metropolitan-scale public and private networks. The topology of these Muni WiFi[†] deployments initially leveraged single-radio, in-band, multihop mesh connections between multiple WiFi access points to provide interconnection backhaul between each node and the operator's points of interconnection to the Internet or private intranet (Fig. 7.1).

As operators gained experience with these Muni WiFi deployments, it became apparent that the single-channel mesh-routing algorithms were essentially self-jamming and significantly constrained the access capacity and available data rate to users on the network. In response to these limitations of the single-radio access point solutions, a number of manufacturers responded by developing multiradio platforms that allowed the WiFi network to allocate all of the 2.4-GHz spectrum in the 802.11(b) and (g) bands to provide user access to the network. The interconnection and mesh backhaul traffic was migrated onto separate radios using the 802.11(a) standard operating in the 5.7-GHz bands (Fig. 7.2).

Metropolitan WiFi networks have been proliferating rapidly, and so is the concept of public–private cooperation between municipalities and WISPs seeking to capture exclusive franchises in partnership with specific cities. In addition, WISPs seeking to provide broadband services to their customers without having to resell telephone company DSL access are driving these new WiFi network deployments, which do not require them to possess licensed spectrum.

---

*HyperLAN spectrum and UNII, Unlicensed National Information Infrastructure, are the license-exempt bands at 5 GHz in the United States and its European Union counterparts.

†Muni WiFi—Municipal WiFi describing what are typically metropolitan area mesh networks, most often developed as either a franchise or public–private partnership.

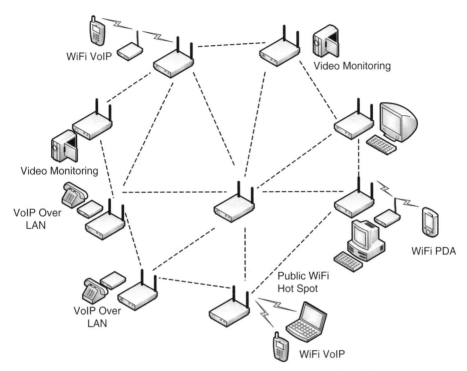

Figure 7.1. Single-radio WiFi mesh, all nodes peer. *Source*: http://www.octoscope. com/images/mimo_10.jpg.

Over the summer of 2007, the bloom came off the rose regarding industry-wide infatuation with municipal WiFi networks that were to be constructed under franchise license agreements and public–private partnership arrangements in several major urban cities. EarthLink, one of the initial leaders in the deployment of metropolitan-scale WiFi networks, came to the realization that adoption of broadband wireless subscription services was far below its forecast uptake of the required number of paying subscribers to meet its original business plan. In several jurisdictions, EarthLink has slowed the expansion of its WiFi network deployments and in pending situations is actively seeking to renegotiate the financial commitments by the municipalities to either partially or fully offset the capital expense of these metropolitan-scale WiFi networks. The initial enthusiasm for municipal WiFi deployments led to numerous public tenders seeking operators willing to bear the cost of building and operating metropolitan WiFi infrastructure under what were essentially exclusive franchise agreements. The fatal flaw in this approach was the typical belief that the cities could benefit without the need for them to make significant capital contributions, or contract for take or pay for service agreements. Despite early deployments coming on-line, the adoption of paying subscribers has to date been insufficient to meet the minimum financial return requirements of investors or operators.

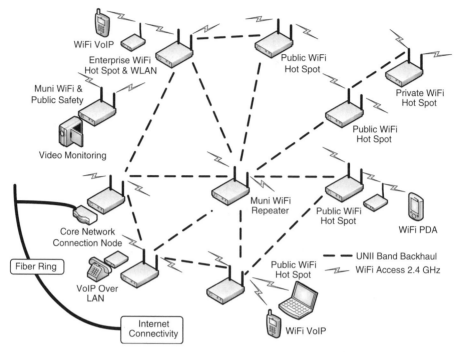

Figure 7.2. Structured WiFi multiradio mesh.

In addition, other municipal WiFi companies have announced plans to either scale back or renegotiate the financial terms of their contracts with various municipalities. MetroFi has been positioning its municipal WiFi business as an advertising-supported system, which is free to all comers. MetroFi has also reached the point where it has determined that there is insufficient present utilization of its networks to generate enough advertising revenue to justify the capital expense required to deploy and manage these networks.

AT&T has also recently been in the process of reevaluating its approach to municipal WiFi markets either as an independent service provider or under various forms of public–private partnerships. It is noteworthy to mention that among the major service providers participating in the municipal WiFi market, AT&T consistently attempted to negotiate take-or-pay contracts with the various jurisdictions it was serving in order to ensure the viability, sustainability, and profitability of its municipal WiFi networks.

ORGANIC WiFi NETWORKS

A large number of citizen-based initiatives are currently under way to create municipal WiFi access networks using the ad hoc deployment of member-owned access points to create an organic tapestry of wireless broadband access coverage.

Most of these initiatives have been fairly small scale, often endorsed and blessed by local municipalities but seldom subsidized or compensated for by the local governments. Many of these initiatives are led by computing enthusiasts who are leveraging such technologies as shareware and donated equipment from either local distributors or major manufacturers, who are trying to gain a position of some dominance in these local markets, should they achieve critical mass and become commercially viable. Leading examples have appeared in San Antonio, Texas, and Seattle, Washington.

One of the leading hardware solutions that is fueling the growth of these organic networks is Meraki Networks, Inc., which was founded as a result of a Massachusetts Institute of Technology (MIT) Ph.D. research project that provided wireless access to graduate students. The Meraki solution has spread into organic network environments in over 25 countries around the world primarily through low cost Web-based direct marketing and through word of mouth. It is a low cost pure mesh WiFi single-radio solution.

In addition to organic, volunteer bandwidth–sharing initiatives, some alternative business models have begun to gain momentum. FON of Spain has been selling consumer-owned shared access WiFi hot spots to a revenue-sharing community of DSL or cable modem –interconnected broadband users, who thereby extend the reach of their private broadband access into the shared access realm of the FON members, called "Foneros." FON has been distributing hot spots to thousands of persons who agree to attach them to their existing broadband connections and allow others to gain access to the Internet under a 50% revenue-sharing scheme with the hot spot owners, which is managed by FON. Recently, the concept gained substantial visibility when British Telecom agreed to open its network to FON users to sell its fallow capacity to FON roamers. In many jurisdictions, there are legal exposures still to be resolved that restrict retail customers of cable or telephone companies from reselling or extending open access to their private Internet connections. The underlying Internet access providers, who ultimately control this space, will have to determine if these schemes are to be embraced as a means of increasing the reach of their subscriber base or if it is completely an anathema to their core business objectives (Fig. 7.3).

In addition to municipal WiFi networks, a growing number of major universities and smaller colleges and universities have deployed campuswide WiFi or, at a minimum, hot spots in select high density areas of the campus to provide students with broadband access regardless of their location on campus.

The FCC has been extremely supportive of these license-exempt operations through its allocation of substantial amounts of spectrum to support all of the expansive applications enabled by WiFi technology, including consumer, commercial, and public safety. With the widespread proliferation of WiFi technology access across various industry segments, the FCC has allocated 50 MHz of bandwidth exclusively to public safety applications, responding to requests from the public sector. The advent of the multiradio WiFi access points has allowed the seamless incorporation of the 4.9-GHz spectrum into public–private metropolitan networks, delivering significant public benefits to participating municipalities.

Figure 7.3. FON WiFi shared access hot spot.

PUBLIC SAFETY WiFi DERIVATIVE

The FCC allocated 50 MHz of spectrum in the 4940–4990 MHz band (4.9-GHz band) for fixed and mobile wireless services dedicated for use by public safety agencies. This allocation will provide public safety users with additional spectrum to implement new broadband applications, including (1) wireless local area networks for incident scene management; (2) mobile vehicle and personal communications; (3) surveillance video cameras; (4) digital dispatch operations; (5) real-time field access for suspect or missing person image files; and (6) site maps, building blueprints, and hazardous materials records for incident locations.

LICENSE-EXEMPT SPECTRUM

Earlier the FCC had dramatically expanded the pool of spectrum allocated for license-exempt operations. The concept of shared spectrum with no formal rights other than users agreeing to comply with technical service rules is designed to minimize interference with other operators and to accept co-channel interference from other operators in the band. Traditionally, these bands were created to support products that required low power radio frequency capabilities for a wide range of industrial, scientific, and medical (ISM) applications. Spectrum has been allocated throughout the frequency table to support ISM requirements for frequencies ranging from below 100 MHz to over 100 GHz. Figure 7.4 presents the ISM bands at 2.4 GHz.

The 802.11 standard defines 14 fixed 22-MHz-wide channels in the 2.4-GHz ISM band. In the United States, only the first 11 channels of the WiFi standard are available for use under FCC rules. Only three channels (1, 6, and 11) can be used at any single location in order to avoid co-channel interference, which will raise the noise floor for the spread spectrum systems operating on any of the overlapping channels. Overlapping service users will result in reduced data throughput, and performance will degrade because of overlapping channel size and interfering signal strength.

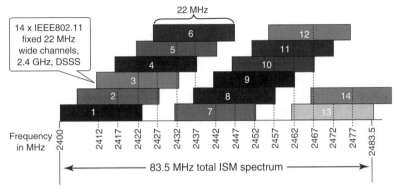

Figure 7.4. ISM bands at 2.4 GHz.

These are the frequencies that must support literally all of the 802.11(b)- and (g)-originated traffic from all sources. The broadband channel frequencies are extremely few in number, and the rules for license-exempt use have no legal recourse to interference.

UNII Bands (United States) and HyperLAN Bands (EU) at 5 GHz

The upper three UNII bands are authorized to operate at the higher power output required for outdoor use, 5.15–5.35 GHz (operating at ≤250 mW) and 5.725–5.825 GHz (≤1 W) frequency bands, and are designed to allow for higher data rates (up to 54 Mbps) under the IEEE 802.11(a) standard, which allocates 12 fixed bandwidth nonoverlapping channels for higher power applications (Fig. 7.5).

THE COORDINATED "SHARED COMMONS"

In April 2007, the FCC released its final report and order on the establishment of a new type of semilicensed spectrum in the 3.65 to 3.70 GHz bands. The service rules for this spectrum are unique in that licensees must obtain a nationwide permit to operate in the band from the FCC, which then allows them to construct a radio system infrastructure at any location nationwide. Once constructed, these new facilities are protected from subsequent deployments in the same area on a first-come, first-served basis. The lower 25 MHz of bandwidth is ordained for formal network coordination using the inherent capabilities in the IEEE 802.16(d) WiMAX and other contention protocols that can prevent interference only with other devices using the same or similar protocols or standard; this will require coordination of network clocks by all operators sharing the bands. The full 50 MHz is destined to be open to equipment that is capable of tolerating and preventing interference with other dissimilar contention technologies.

Thus, a second operator who constructs facilities in the same area must coordinate with the existing operator, who is the first, and all subsequent

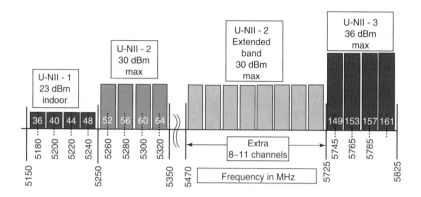

Note: 12 nonoverlapping channels within U-NII-1, U-NII 2 and U-NII 3.

Figure 7.5.   UNII frequency tables and power levels.

operators in these frequencies in a given area must agree to coordinate their operations in descending order of preference with operators who were there first.

The FCC has mandated a forced sharing protocol that allows multiple operators in the same service area to coordinate their network operations that can make use of the unique capabilities of the 802.16(e) standard, i.e., the technology adopted for Mobile WiMAX. This protocol allows for the synchronization of multiple systems sharing similar spectrum. Time slot synchronization, adaptive power and modulation synchronization, and frequency use and release synchronization are all available in real time to allow multiple system operators to seamlessly coordinate their systems.

As appropriate equipment is delivered to the market, this new band will provide a unique opportunity to examine the merits of spectrum allocation, which rewards early construction facilities, versus the present approach to auctioning spectrum, which often results in the warehousing of precious spectrum assets rather than seeing it used to deliver new and innovative services. This approach is a grand experiment in spectrum allocation policy that is almost purely market based. Our industry will be watching closely to see if this approach is worthy of further extension into other spectrum bands and applications.

THE WiFi ALLIANCE*

The WiFi industry owes much of its success to the efforts of the WiFi Alliance. This trade organization was established to ensure the seamless interoperability of devices built to the various 802.11 specifications. All equipment suitable to display the WiFi-certified logo has been tested to meet interoperability standards established and managed by the alliance.

---

* The WiFi™ symbol are the property of the WiFi Alliance.

The WiFi Alliance was founded in 1999 to certify WLAN equipment, but has expanded its influence across the entire spectrum of WiFi-enabled devices, which now range from municipal access network equipment to cordless WiFi-enabled phones.

Over 300 member companies, representing over 20 countries, now participate in the alliance. The tremendous success of WiFi across global boundaries has spawned the creation of a parallel organization to replicate the formula in the 802.16 standards domain. The new organization is called the WiMAX Forum and has grown dramatically over the past six years.

WiMAX FORUM*

The WiMAX Forum was originally founded in 2001 to address issues of interoperability for point-to-multipoint microwave radio systems above 10 GHz. It now has over 470 members from all corners of the wireless communications industry. Subsequently, it was reorganized to concentrate on the working group for such systems operating below 10 GHz, known as 802.16(a) and 802.16(e). The WiMAX Forum was established as a not-for-profit industry organization dedicated to the promotion of WiMAX interoperability testing and certification for all 802.16 (IEEE), the Korean WiBro, and European (ETSI) HyperMAN standards–compliant equipment.

During the process of developing the standards for point-to-multipoint fixed services, several members of the working group began to advocate for an enhanced version of the standard to address broadband mobile wireless services. This bifurcation in the working group spawned the creation of the 802.16(e) working group. Adding to the already extensive complexity and confusion throughout the wireless industry over the WiMAX "standard" is the fact that these two standards are neither interoperable nor technically similar. The 802.16(a) standard was finalized and uses OFDM/TDMA architecture to provide point-to-multipoint fixed services. The 802.16(e) standard uses OFDMA technology at its core and has been designed to support full mobility services. The driver for these new metropolitan wireless standards was the enormous success enjoyed by WiFi, a far simpler and less powerful technology platform.

Historically, standards were of little practical value in point-to-point microwave systems, as it was logical to use the same vendor for both ends of a communications link. However, as point-to-multipoint systems began to emerge as broadband wireless solutions, there have been powerful business incentives for service providers to be able to source user terminals from multiple vendors. Further, the widespread proliferation of compatible user equipment through the incorporation of compatible chip sets in laptop computers, handheld computers, and mobile telephones provides a market lever that leads to the potential for rapid market expansion by service providers deploying compatible infrastructure. How the WiMAX standard fares in the market is still to be determined, but with the

---

*The WiMAX™ symbol is the property of the WiMAX Forum.

enormous sponsorship behind the forum and the pioneering efforts of Sprint Nextel and Clearwire to implement large-scale Mobile WiMAX networks in the immediate future, the industry will have its first substantive opportunity to examine the future for 4G wireless networks, featuring true broadband to individual users, IP-routed packet data architectures, full high-speed mobility, and open platforms for user-determined software and applications.

8

ALWAYS BEST CONNECTED

PRODUCT DEFINITION FOR BROADBAND WIRELESS SYSTEMS

Always Best Connected [1] defines the key drivers that have emerged which provide significant support and influence to the wireless broadband business opportunity. These include several progressive developments in the realms of regulatory, technology, standards, financial, and customer demand. Ultimately, these drivers all focus on the demand side—from communications services end users—ranging from the high value 24/7 mobile individual to public safety agencies seeking to upgrade their obsolete networks to comport with the broadband information society. The new network paradigm requires the full suite of data communications services to be available, whenever and wherever the need to communicate arises, using low cost, multiprotocol devices—at affordable prices, supporting users with the ability to make choices. In short, the requirement is to be *always best connected.*

One of the leading research reports focused on user requirements is that of the working group on "The Operators' Vision for Networks Beyond 3G" Eurescom report, May 2003 [2]. This report is of particular interest as it provides concrete insight into the strategic direction of integrated network operators at the cutting edge of technology (Fig. 8.1).

The main point made by the operators studied in the EURESCOM report is that end users (their customers) do not care about underlying network technology specifics as long as the network works, is simple to use, and is able to be

Wireless Broadband. By Vern Fotheringham and Chetan Sharma
Copyright © 2008 the Institute of Electrical and Electronics Engineering, Inc.

Vision	Functional Requirements	Operator Requirements
It Works	Sufficient bit rate for true broadband services Seamless interworking between access networks	Implementation of access networks for hot spots, nationwide coverage and special purpose networks. Network backbone capable of handling internetworking through media gateways
It's Simple	Plug & play user self installation New service discovery network functionality Unified access from different terminals and multiple networks	Middleware for service discovery Common service platform for different networks
It's Personalized	Intelligent mobile agents Virtual home environment Context aware applications	Access to home domain service platform from anywhere

Figure 8.1. Customer requirements for next-generation wireless broadband networks.

personalized to their specific needs. These end-user requirements for services set the stage for network operator requirements regarding the types and kinds of enabling technologies they will be seeking from solution vendors.

The technology and product development engines fuelled by the demand side represent different approaches to shaping the market. The following section reviews each of the primary engines that will drive the market. Ultimately, it is the delivery of reliable, cost-effective services with a high level of utility for the users that will define the winners and losers of the broadband wireless future.

TECHNOLOGY DRIVERS

Network Convergence

The main factors at work are related to a phenomenon known as "network convergence," arising from the observation that the developed societies of the world had separate communications networks for delivering telecommunications services—broadcasting, voice, and data—to the population. Along the way, we added coaxial cable and satellite systems to augment TV broadcasting, and cellular mobile networks to extend the copper-based, circuit-switched voice infrastructure. These application-specific networks duplicated services and required substantial amounts of capital to reach an increasingly fragmented market. A key driver of convergence is the growing "cost of ownership" for infrastructure, which must pass all the potential subscriber locations before capturing only a percentage of the "homes passed" as paying customers. The potential to conserve capital by deploying a single pure IP converged physical network that service providers can use for delivering any mix of services or that could be shared among multiple service providers will allow societies to improve their communications capabilities while retaining fiscal resources that can be used to enhance other sectors of their economy.

A major constraint imposed on the network convergence process by legacy network stakeholders is their desire to "protect investments" already made. It is important for the purpose of understanding the present situation to realize that no "disruptive technology" ever makes it through the gates of the information society without leveraging existing infrastructure, and making most targeted stakeholders better off in all material respects. Technological leadership is always up for grabs, but new technology leaders must find ways to provide tangible near-term financial benefits to peers, regulators, customers, and end users to ensure their own success.

Mobile IP with Quality of Service

To achieve the desired network convergence, it is necessary to separate the network into layers and elements to enable the convergence process to be managed in steps. The ability to manage the QoS over shared packet networks is essential to meeting the expectations of customers. The industry thus far has done a fairly poor job of delivering QoS in the delivery of data services, although the reliability of the voice services in the developed nations now approaches that of landline telephone service. Our global introduction of true broadband services as we move into the 4G era will come in parallel with the integration of strong QoS capabilities in the new broadband wireless networks.

The apparent chaotic torrent of wireless products and solutions flooding the market every day represents all the competitive offerings that conform to the standards established for the purpose of achieving the goal of the giant universal project supported by the Global Information Society: Network Convergence. The first level of convergence is achieved when the predominant data network protocol—IP—is adopted across network platforms. This allows the end user to install software, such as browsers, that communicates using IP on his or her device, and so the user is indifferent about the underlying physical network. The user's only interest is that the access network recognizes him or her with a single identity, offering the best communications price/performance, and that the user can enjoy secure network access wherever he or she goes. This has the effect of turning the closed vertical networks into an apparently open and horizontal architecture. In practice, a single identity across multiple physical IP networks requires terminal middleware that is network aware and allows secure login and seamless network handoffs.

Wireless Technology

Beyond network convergence, the increasing importance of wireless technology is the dominating theme of the next era of telecommunications services. Standardization and interoperability are key issues behind the success of wireless technology in terms of unit sales. Subscriber figures show that there are now more wireless subscribers than fixed-line subscribers in the world. Total subscriber figures are approximately 1.2 billion GSM users and 200 million CDMA users. It is useful to remember that the laws of physics will always impact the structure of the network. Radio signals are strong near the transmitter and will eventually become weaker with distance, given a certain frequency and bandwidth. The unique benefits of OFDMA technology lie with improving the overall network performance and efficiency of wireless networks. Orthogonal frequency division multiple access Mobile WiMAX systems have demonstrated increases in spectral efficiency on the order of four times over CDMA networks operating in equivalent amounts of spectrum.

Personal Area Networks

The first level of the overlapping network standards (Fig. 8.2) is the wireless personal area network (PAN), which is used by individuals to communicate with their personal devices, and allowing communications between the personal devices of others in close proximity. Key benefits to the user are ease of use and freedom to move. In PANs, the average distance between transceivers is likely to be measured

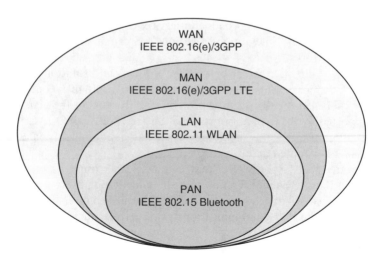

Figure 8.2. Overlapping global wireless access standards.

in meters. These networks usually have gateways to the local area network (LAN) of the user in their home or enterprise.

Local Area Networks

Local area networks are deployed to serve a specific user group located in reasonable proximity, such as in an enterprise or a home. Such affiliated individuals share data and program files and the gateway providing access to external networks. Beyond the family or workgroup, a temporarily shared location as an affiliation criterion has led to the emergence of public wireless LANs (PW-LANs or "hot spots") in locations where individuals are by choice or necessity in a stationary mode, such as in hotels, waiting rooms, and at food and beverage establishments, and while on board shared means of transportation, such as trains, ships, and airplanes. The main benefit to users of such hot spots have so far been to have access to a shared-cost Internet gateway to public and private network resources. The average distance between user transceivers and wireless LAN access is typically measured in tens of meters.

Metropolitan and Wide Area Networks

Mobile communications network access serves two distinct environments: (1) the metropolitan area network (MAN) and (2) the wide area network (WAN). The MAN is a special case of a WAN in that the resource cost-sharing opportunity is much higher where there are high concentrations of people. This opportunity means that network access can be economically provided with a higher network granularity, but it also becomes a cost-driving service requirement due to the presence of buildings or natural obstructions that place requirements on signal strength, which can be very diverse between users, depending on their physical position, such as being indoors within specific buildings or outside in a wide open space. In MAN, the distance between transceivers typically averages between 500 meters and a few miles. In cellular networks, the return path from the weak transmitter of the small battery-powered user device is the weakest link in the communications path. In the wide area, away from the metropolis, the economics of network buildout changes dramatically. The number of simultaneous users per network base station, or per square kilometer, falls dramatically, while the added issue of transient users and their propensity to amass in large numbers in the early morning and early evening near the metropolis and on weekends in certain attractive locations add to the challenges of network capacity planning. The average distance between WAN transceivers may be measured in several to even tens of kilometers or miles if the user devices are provisioned with high gain antennas.

Standards Overlap: Ethernet/TCP/IP

Weaving together each of these "area-specific" domains is the unifying power of IP technology. The technologies that have been developed to address the unique requirements of a particular set of application and area-specific characteristics share the standards platform of IP to simultaneously address each of the PAN-LAN-MAN-WAN-SAT classifications.

The main trade-off from the user's point of view is between transmission speed and distance. For the individual user, practical issues such as device versatility, access to broadband services, and devices at competitive costs are important additional considerations. Few people will drag five specialized communications devices around. The industry has experienced multiple approaches to dealing with different business models: specialized devices for given situations, best all-round solution, multimode devices, and the increasingly important embedded solution. Making communications part of other objects that individuals own or use will address the objection of adding additional items to the briefcase, backpack, purse, or pocket. Mobile communications technology is being engineered into clothes, cars, personal computers (PCs), and PDAs, which allows the user high degrees of personal freedom and the vendor a high volume opportunity with new product design challenges and opportunities. The embedded solutions give rise to the importance of the PAN in any product line. The emergence of PAN-based solutions may impose important feature requirements on other communications technology products and give rise to a new product discriminator on the end-user side.

EVOLVING WIRELESS BROADBAND MARKET SEGMENTS

- Mobile system operators/Cellular ≫ Vox to converged VoIP/data
- Internet service provider ≫ Wireless Internet service provider
- Competitive local exchange carrier ≫ Special access/telco bypass/metro Ethernet
- Incumbent local exchange carrier ≫ Fixed mobile convergence/nomadic/ mobility if no cellular operation
- Mobile system operators ≫ Underlay/overlay BBWA upgrades
- Cable ≫ Nomadic/mobility service expansion
- Fixed wireless access ≫ Interconnection and mobile backhaul/nomadic/ mobility service expansion
- Interexchange carrier ≫ Converged services migration
- Public safety ≫ Federal, state, county, municipal interoperability
- Internet portal operators and content hosts ≫ Advertising–supported mobile services

The potential wireless broadband market stratifies into nine segments.

1. Mobile network operators/"Cellcos" exist around the world in practically every country. It is a clearly defined operator group, and the members are readily identifiable. Mobile network operators have a clearly defined need

for network solutions supporting higher speed Internet access. These operators are likely to have cell site infrastructure in place that will make this group potentially the lowest cost provider of additional, higher speed broadband wireless services from a total cost of ownership perspective.

2. Incumbent local exchange carriers are focused on their copper and fiber assets, providing xDSL services. These ILEC-based DSL operations have led the high speed access services mass market. The lack of a mobility service platform is driving many of these legacy operators (who in many cases have previously spun off their cellular assets) into the WiMAX/ WiBro market.

3. Competitive local exchange carriers increasingly look for wireless technologies to provide high speed access competing with the copper-based solutions of the ILECs. These wireless technologies would cut them free from the provisioning process of the ILECs. In the United States recent court decisions have essentially ended the Unbundled Network Element rate caps and forced CLECs into bilateral negotiations with each ILEC they had previously relied on for their special access circuits to connect their business customers. Broadband wireless will gain a substantial boost from this public policy shift, now providing significant incentives for CLECs to build autonomous bypass access networks.

4. Wireless Internet service providers are facing the same challenges that CLECs endure with resold access from incumbent telephone companies. In second- and third-tier markets the use of license-exempt spectrum will be a significant contributor to allowing WISPs to gain independence from the ILECs; the potential for WiMAX in these markets is especially promising.

5. Cable TV operators are rapidly leveraging the bandwidth advantage of their coaxial networks to deploy cable modem access for high speed Internet services. Cable companies most often lack any mobile or portable products to bundle with their fixed voice, data, and video converged services. The lack of mobility services will become a growing strategic disadvantage for cable operators, and it is anticipated that they will actively participate in the acquisition of spectrum assets.

6. Fixed wireless access providers in both the licensed and license-exempt bands are numerous, but with only modest penetration of their still evolving business models. This group is likely to embrace WiFi and WiMAX systems for deployment in unlicensed frequency bands to improve network reach, to get more subscribers under their footprint, and to expand services offerings on the basis of multimode terminals. This possibility is likely to lead these groups, including hot-spot operators, to develop more substantial relationships with their customer base, offering home, office, and hot-spot WLAN services. They could find themselves in head-on competition with mobile operators, whether or not they go for the provision of mobile services.

7. Spectrum holders that own fallow spectrum in several bands with aspirations of becoming mobile service providers are still waiting for the Mobile WiMAX standard to be commercialized on a wide scale. These include companies with licenses in the Personal Communications Service (PCS), Wireless Communications Service (WCS), Broadband Radio Service (BRS), Educational Broadband Service (EBS), 800-900 MHz cellular, Mobile Satellite Service (MSS) and other licensed spectrum below 3 GHz.

8. Public safety applications are gaining in importance. Recent allocations of radio spectrum exclusively to the public safety community in the 700 MHz and 4.9 GHz bands are driving substantial opportunity for manufacturers and system integrators to extend 4G-based services into the public safety market.

9. Internet portal operators such as Google, Yahoo, AOL, and MSN are all trying to create the new advertising-supported business models that will mark next-generation wireless systems, recently named "Sellcos."

OPEN SYSTEMS AND INTELLIGENCE AT THE EDGE

Layered Networks and Interconnected IP "Clouds"

The wireless broadband extension to the Internet will continue the trend of moving intelligence to the edge of the network. As the power increases in our user devices, the ability to store locally enormous amounts of information, which was historically stored in databases that were attached to the network in or near the core, also increases. Given the multigigabit storage capacities of next-generation handsets, much of the information that was historically stored online will now be available in the handset. Content of all types (video, data, and audio files) will be cached in edge devices. Traditional routing and network access security settings and other network functions will also find their way into the edge device. The concept of "ditch the switch," wherein network control features will move from centralized control computers and network switches to the edge devices, will come into effect. The resulting shift in control from a long list of sticky entanglements that historically existed in the relationship between the service provider and the subscriber will begin to diminish as control over the relationship between service provider and customer tilts toward the consumer.

Stratified Network Topologies

The migration to IP networks between and among so many separate service providers that will inhabit the new world of interconnected packet networks will result in complex layers of stratified networks, which will ultimately allow subscribers to seamlessly move between and among them.

As the legacy telephone companies complete their ultimate migrations to next-generation networks, and as cable and cellular network operators extend their franchises deeper into the broadband Internet, the routing of traffic across end-to-end packet networks will result in the creation of numerous new inter-mediation opportunities for intercarrier settlements, traffic accounting and usage audit, roaming, and digital rights management payment settlements.

Wireless broadband networks will also stratify into overlapping layers of coverage and capacity, with the wide-area coverage and vehicular mobile services to be provided over a macrocellular overlay. In-building penetration and higher data rate services will most often flow over microcells supporting service radii from the base stations of about 500 meters. Deep in-building penetration will often be delivered using pico cells and fempto cells that augment the network capacity by keeping the cell site airtime on alternative infrastructure, typically interconnected using alternative media, such as cable modems or DSL IP connections. In addition to these underlay networks, there are also emerging distributed antenna solutions that extend the reach of selected cell sites using remotely located antennas that leverage the cell site electronics. An alternative approach to extending the reach and especially to improving the performance of the low power return path from handsets is the addition of wireless in-band repeaters. Wireless repeaters have been developed in configurations for both outdoor and indoor deployments.

It is too soon to tell the overall impact of native protocol fempto cells operating in the fixed mobile convergence space, as opposed to what will inevitably be low power versions of WiMAX technology. The dual-mode hybridization of WiMAX and WiFi also holds promise for leveraging the deep penetration of WiFi access points in both public and private environments. It is virtually inevitable that early Mobile WiMAX–enabled handsets will also incorporate WiFi dual mode capabilities.

How the HSPA and advanced versions of EV-DO extend into the native protocol fempto cells to capture a larger role for legacy cellular operators in the fixed mobile convergence market will be worthy of close attention by industry planners and investors.

RADIO NETWORK SYSTEM ENGINEERING

Open Broadband Access Network

Representatives from Integrated Services and Networks Operator Telenor of Norway made an important observation regarding convergence in the context of the Open Broadband Access Network (OBAN) project: The emergence of broad-band access is intensifying as the issue of growing cell site interconnection and backhaul capacity requirements intersects the radio access network (RAN) fabric. One approach is to reuse the access network capacity when not tasked with access. The European research project OBAN is looking into opportunities for sharing this access network capacity with mobile networks, as the base stations in mobile networks increase in numbers for every generation. Note that the numbers corresponding to coverage areas are revealing: 1.8 km average radius for GSM, 600 m average radius for UMTS 1, 200 m average radius for UMTS 2, and about 60 m average radius for WiFi (the Telenor proxy for 4G). Thus, the number of base stations required for coverage and increased capacity is growing exponentially, while the cost of the transport network follows. Ultimately, base station hardware is a minor percentage of total cost of ownership of mobile networks. There is a high economic reward for operators that are able to

- increase the deployment radius for mobile wireless broadband, and/or
- separate voice and data, allowing for expensive broadband backhaul to be limited to high capacity utilization
- share (backhaul) infrastructure among operators, and/or
- use the fixed-line access network for backhaul by opening the broadband access network for such dual use (network convergence!).

REFERENCES

1. Johannessen T. Always Best Connected. Blackwolf Consulting.
2. Kellerer W. editor. The operators 'vision of systems beyond 3G: Operators' key drivers for systems beyond 3G. Eurescom Report; May 2003.

9

BROADBAND IP CORE NETWORKS

The arrival of the next-generation network (NGN) IP core packet data network technology has evolved to the point of no return. The migration from the legacy circuit-switched telephone networks to fully routed packet networks leveraging all common off-the-shelf tools and network elements is a foregone conclusion. Network planners at both legacy telecommunications companies and new Greenfield start-ups in every sector of the telecommunications and digital communications industry are committed to IP deployments. The speed of this transition is surprising to even the most sanguine pundits.

The integration of core NGNs as central to the emerging wireless broadband initiatives simplifies and enhances the flexibility for creation and rapid implementation of new services. Next-generation network architectures eliminate the need for layers upon layers of specialized network element controllers. The business benefits dramatically from the reduction in costs, operating expense, and improvement in the total cost of ownership, thus enhancing profitability.

USER AUTHENTICATION AND LOG-IN

AAA: Authentication, Authorization, Accounting

Authentication is the method of identifying every user seeking to log on to the network. Previously, users were often required to enter a user name and password

Wireless Broadband. By Vern Fotheringham and Chetan Sharma
Copyright © 2008 the Institute of Electrical and Electronics Engineering, Inc.

assigned to them. As we move into a period where always-on broadband packet networks are used to deliver services in real time to mobile devices, there is a need for new forms of automated, real-time authentication. It is likely that the various authentication schemes used in the cellular industry, both SIM based, and embedded network user registration databases will also emerge in other domains of the IP centric network world to enable passive authentication.

Authorization controls the rights and privileges available to users seeking access to the system on the basis of their subscription rights. Once the users pass through the authentication process, they are matched to user database information, which defines their service profile and then admits them onto the network and into the service domains that they are qualified to receive.

Accounting provides a record of all aspects of a connected session to the Internet. It provides the documentation needed to drive billing systems and support network audit purposes. Time of the connection initiation, length of the session, time logged off the session, and the number of bits, bytes, and packets passed during the session are all provided by the accounting function of the AAA servicer. Accounting data are also used to support the need for network engineering to run systems analysis reports for trend tracking, capacity planning, fraud detection, billing, auditing, and cost allocation.

RADIUS (*Remote Authentication Dial-In User Service*) is a common platform for implementing AAA services. Originally created as the dial-in authentication system for the Internet by the Internet Engineering Task Force (IETF), RADIUS has subsequently been updated for broadband always-on packet networks. Although it is not a formal standard, the RADIUS specification is supervised and maintained by a standing working group of the IETF. There are numerous vendors for RADIUS servers, ranging from open-sourced shareware to highly redundant and carrier-class servers.

PROVISIONING

Provisioning, the inauguration of new services, upgrades, or the need to manage subscriber moves, adds, and changes, is one of the greatest risks to the future deployments of broadband IP wireless networks as parallel competitors to cellular operators. The new entrants must ensure that they have implemented robust, scalable, reliable, and accurate new account provisioning systems. The seamless integration of automated systems that are linked deep into the network and capable of automatically populating or updating numerous databases simultaneously is an essential capability of future NGN wireless networks.

The ability to leverage user self-installation and self-registration within the system is essential for a successful twenty-first century wireless broadband service provider. As the new alternative to extremely powerful and well-established competitors, the coming wave of new broadband wireless service providers will have to get this right. Failure to do so will result in a botched deployment and market introduction. The old adage "You only get one first impression" is painfully true in the highly competitive wireless industry. The inability to make a positive impression on your customers at every point of intersection will result in lost customers and a failure to maximize the potential of the business. Failure to invest in highly integrated and robust back-office and network support systems under the false impression that it will cost too much or take too long is a too common and foolish rationalization. No manual interim processes requiring forms juggled by clerical employees are sufficient to meet the requirements of any fast growing telecommunications organization of significant scale in the modern era. Time to market is a key strategic imperative in our industry. However, failure to implement a comprehensive suite of Billing and Operational Support Services (BOSS) early will lead to untenable situations that will ultimately require massive amounts of cleanup and management distraction, notwithstanding the likely damage imposed to the reputation of a nascent service provider.

Billing and Operational Support Systems (BOSS)

- Plug and play—user self-installation
- Automated provisioning (BOSS)
- Prepaid and postpaid billing platforms
- Metered services
- Flat rate bucket plan services
- Calling party pays versus called party pays
- Cellular 800 numbers (toll-free airtime, stillborn)

FIXED- AND MOBILE-CONVERGED SERVICES OVER A UNIFIED PACKET NETWORK

IP Multimedia Subsystem

The IP Multimedia Subsystem (IMS) is an architecture design to enable transition of legacy circuit-switched networks to packet-based IP networks through an interoperability regime that allows for a seamless evolution of service delivery to both legacy and NGN customers simultaneously.

IP Multimedia Subsystem is a general-purpose, open industry standard for voice and multimedia communications over packet-based IP networks. A transitional core network technology is based on Session Initiation Protocol (SIP). It provides the core network foundation for services such as VoIP, push-to-talk (PTT), push-to-view, video calling, and video sharing. Existing carriers are leveraging IMS to transition from the circuit-switched core networks of the legacy cellular voice-centric business to a flat IP-based core network, which will support all types and classes of services on a unified network. Greenfield WiMAX network deployments are also leveraging IMS as the core platforms to enable them to spoof all traditional cellular services in an all IP environment, but without having to reinvent the wheel for managing packet service integration.

IP Multimedia Subsystem was originally conceived by the 3GPP with a modest objective to enable GSM cellular operators to deliver IP services over GPRS networks. The 3GPP subsequently expanded the original vision to embrace the 3GPP2 road map for LTE of 3G networks of all types, including CDMA2000, WiFi, WiMAX, and even fixed-line networks. It has also been adopted by Telecoms and Internet Converged Services and Protocols for Advanced Networks (TISPAN), the ETSI standardization body for next-generation networking protocol harmonization. IP Multimedia Subsystem is finding its place as the leading candidate to allow for the synthesis required in the fixed-mobile convergence space (Fig. 9.1).

The challenge that faces the wireless industry following its conversion to a packet-only architecture and broadband services is its present reliance on virtually every value-added service running on discrete network infrastructure dedicated to functioning as completely vertically siloed systems. Over the past several years, the large number of mergers among the cellular carriers have exacerbated the problem, with dozens of server platforms running in widely dispersed regional locations, struggling to just keep up with existing demand, much less gaining on a systemwide integration of fully converged NGN-based broadband packet core networks. The mere challenge of maintaining current cash flows in a highly competitive market weighs heavily against even the most progressive chief technical officers (CTOs) and chief information officers (CIOs) that are trying to position for the coming broadband data wars. A recent example of the seriousness of these technical and market overhangs is the challenges that led to the ouster of Gary Forsee at Sprint-Nextel after the company failed to gain significant market advances post merger. More interesting was the widespread pressure placed on Sprint to abandon or delay the rollout of its nationwide Mobile WiMAX NGN wireless broadband network,

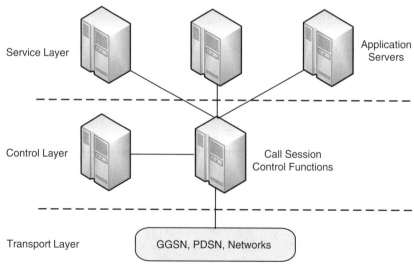

Figure 9.1. IMS-layered network architecture.

because it was too costly, at the expense of its existing 2G and 3G revenue streams. At the time of the final edit of this book we have reached the point of the merger announcement between Sprint Xohm and Clearwire with expansive backing from Comcast, Google, Time Warner, Brighthouse, and Intel.

Network Backhaul and Interconnection

The construction of end-to-end pure IP broadband networks will require new and/or expanded approaches for cell site interconnection, distributed over a wide range of capacities and form factors, including towers, building rooftops, and indoor micro, pico and fempto cells, with numerous permutations anticipated for each of these form factors. The objective of providing sufficient broadband backhaul capacity is a major challenge.

Carriers will seek to connect to fiber appearances whenever possible; however, there will be numerous locations with no fiber in sight. At these locations, the installation of Ethernet backhaul radios to manage the large data flows, which will be aggregated over the new 4G wireless networks, is the growing solution of choice. The legacy cellular operators have long leveraged the copper network to interconnect the heavy concentration of voice traffic in 2G and 3G cellular networks. Typically, a cell site would have been connected with two to four DS-1 private line circuits to support the backhaul of what was predominantly narrowband voice traffic. This is only about 6 Mbps and will prove to be completely inadequate in managing the estimated 30–200 Mbps of the backhaul requirements that will be needed by the new 3G and 4G wireless broadband networks.

FIXED BROADBAND WIRELESS NETWORKS

The use of Ethernet microwave radio technology for connecting cell sites has been expanding in recent years. Both Fiber Tower and Nextlink Wireless are actively leveraging their extensive millimetric spectrum holdings nationwide to provide licensed point-to-point and point-to-multipoint broadband-fixed microwave services to the expanding need of the cellular industry as they move into widespread deployment of their 3G networks.

The need for additional capacity to accommodate the aggregated mobile data throughput from their HSPA or EV-DO services is growing rapidly. This increased capacity requirement will continue to grow and will be exacerbated by the deployment of Mobile WiMAX and GSM LTE wireless broadband 4G networks. The amount of capacity that is required to interconnect these distributed broadband wireless access nodes is substantial. Envision a cell site with four 90° sectors, with each sector transmitting and receiving in excess of 50 Mbps, and then as we combine the data streams from each of the four sectors, the combined capacity requirement swells to approximately 200 Mbps. There are efficiency techniques that can allow for a backhaul that is less than the entire gross combined data rates; however, to enjoy the highest quality of service and least amount of latency, overbooking should be kept to a minimum. With this in mind, there is need for fractional GigE wireless connections; 100 Mbps Ethernet radios can be used to serve the first wave of 4G network deployments, but this will only be a stopgap measure for a few years at best.

Fixed Wireless Spectrum Availability

Numerous spectrum bands may hold the promise of sufficient capacity and bandwidth to enable multihundred megabits per second of capacity over millimetric microwave links.

Below are listed some of the license-exempt or licensed bands that are either in private hands or are available to be licensed on a link-specific basis.

ISM Bands (900 MHz, 2.4 GHz). The most often used ISM license-exempt, or often referred to as "unlicensed," spectrum are the 902- to 928-MHz and 2400- to 2500-MHz bands (see Fig. 7.4). There are a number of other ISM bands that have historically been used for narrowband services, such as garage door openers, analog cordless phones, and security systems in HF and VHF bands. In addition, there are allocations at 24, 61, 122, and 244 GHz that have been established for broadband applications, including vehicular radar and wireless local area networks or data bridges. Specific ISM frequencies in the millimetric bands include

- 24–24.25 GHz
- 61–61.5 GHz
- 122–123 GHz
- 244–246 GHz

UNII Bands (5 GHz). There has been a lot of progress in the creation of near-line-of-sight OFDM-based point-to-point microwave radios in the license-exempt UNII bands at 5 GHz (see Fig. 7.5). Presently, because of limited bandwidth and power constraints common in the UNII bands, the data throughput capacity of most solutions will be spoofing 100 base T (bT) at about 70-Mbps TDD. Frequencies include the following bands:

5.150–5.250 GHz (23 dBm indoor)

5.250–5.350 GHz (30 dBm outdoor)

5.470–5.725 GHz (30 dBm outdoor extension band)

5.725–5.875 GHz (36 dBm high power outdoor)

Point-to-Point Common Carrier Microwave Bands (10, 18, 23 GHz). The traditional point-to-point link licensed FCC Part 100 common carrier bands are available over many path routes, although they reflect band plans that typically reflect legacy telephone company circuit requirements for multiple DS-1 (N × 1.544 Mbps) and DS-3 (45 Mbps) capacities. Most common carrier link licenses are granted use of FDD band plans with insufficient contiguous spectrum to support data rates in excess of multihundred megabits per second.

Point-to-Point Applications and Multiple Point-to-Point Deployments

Microwave radio systems have evolved from their roots as point-to-point long-distance transport for voice telephone traffic. Some of the beautiful and massive installations of the old Bell System's long-distance microwave network with large cornucopia (or "Horn of Plenty") antennas, mounted on tall tower structures, mountaintops, or telecom-switching facilities, are still a common sight.

Microwave technology has advanced from its roots as an analog radio technology that required hand tuning of individually matched radio pairs and intensive maintenance oversight postinstallation to ensure reliability to the current era where devices can be quickly and easily installed, with any two devices out of production able to communicate and automatically configure themselves for service.

There has been a boon in technology advances in this space since the mid-1980s built on the back of government-sponsored defense industry investments in millimeter-wave radio technologies, originally driven by the "Star Wars" programs in basic research. The dramatic advances in performance and reductions in cost made possible by these industrial policies fed directly into the applied materials sciences for the development of components that can function with the stability required to commercialize products operating at billions of cycles per second. Cost reductions for these devices have been on a continuous curve following the mass-market deployment of DBS receivers and the semiconductor industry's efforts to apply Moore's law for creating microprocessors as both ASIC and generic, high performance devices that can synthesize in software many of the functions that have historically required discrete components via DSP.

Point-to-point microwave links have found common use in private telecommunications networks and are rapidly growing in numbers to provide the ever-increasing demand for the added bandwidth required by 3G and 4G cellular communications systems. As mobile broadband data applications expand, the need to upgrade the amount of bandwidth available at cell sites will become a universal challenge for network operators. Given the lack of widely dispersed fiber-optic network facilities, especially at distributed cell site locations, and the inadequacy of the legacy copper networks to meet the distance and broadband capacity requirements of the near future, it is inevitable that the point-to-point microwave solutions will continue to expand rapidly (Fig. 9.2).

Point-to-point microwave has a number of unique performance and cost advantages for the interconnection of cellular telephone sites with its network control and switching facilities. Unique to the microwave solution is its ability to deliver bandwidth capacity greater than 100 Mbps to existing towers, which can be easily installed at price points that will allow cost recovery in less than one year. The bandwidth delivered can be incrementally scaled up to meet staged requirements of network broadband evolution.

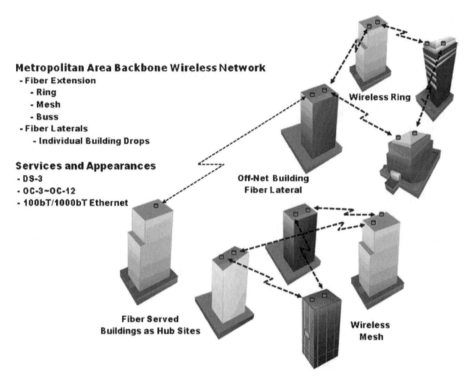

Metropolitan Area Backbone Wireless Network
- Fiber Extension
   - Ring
   - Mesh
   - Buss
- Fiber Laterals
   - Individual Building Drops

Services and Appearances
- DS-3
- OC-3~OC-12
- 100bT/1000bT Ethernet

Wireless Ring

Off-Net Building
Fiber Lateral

Fiber Served
Buildings as Hub Sites

Wireless
Mesh

Figure 9.2.   Point-to-point microwave.

In the United States, the first wave of cellular infrastructure was mostly (about 80%) interconnected using copper DS-1 facilities acquired from local telephone network operators. In Europe and much of the rest of the world, cellular interconnection via microwave was the predominant approach. We anticipate that the use of microwave facilities is going to expand dramatically throughout the United States as true broadband mobile services proliferate (Fig. 9.3).

Point-to-Multipoint Millimetric Microwave Bands

A number of broadband wireless area licenses issued over the past 15 years have to date not found a workable business formula that balances the cost of network and user equipment and the gatekeeper obstacles of roof rights and landlord expenses for riser access and installation against the demand for the amount of broadband capacity that is unique to these frequency bands. To date, the enterprise access markets dominated by DS-1/E-1 connections have sourced primarily from the legacy telephone companies. As the market demand for broadband network access matures to require 10 bT (10 Mbps), 100 bT ($\leq$100 Mbps), and gigabit Ethernet ($\leq$1000 Mbps) services, we anticipate that the millimetric spectrum will finally take its logical place in the stratified and hybrid networks of the future.

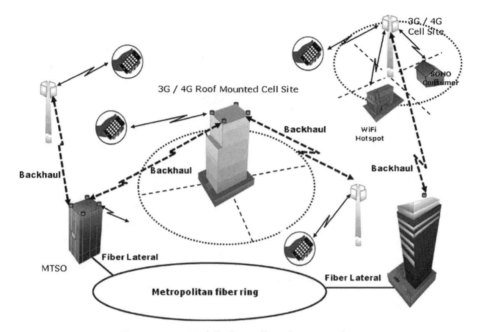

Figure 9.3. Mobile broadband connection.

Figure 9.4 illustrates the various aspects of hybrid fiber and wireless network architectures that will emerge to leverage the unique capabilities inherent in the point-to-multipoint millimetric microwave services.

Local Area Multipoint Distribution System (28 and 31 GHz)

These area licenses exist nationwide, with Nextlink Wireless, the subsidiary of XO Communications, controlling the vast majority of the major population centers. The band has one of the largest single blocks of contiguous spectrum, 850 MHz of bandwidth.

Digital Electronic Messaging System (24 GHz)

The DEMS licenses were acquired by Fiber Tower in its roll-up of the former Teligent and Advanced Radio Telecom spectrum assets. It has 200–400 MHz of bandwidth essentially nationwide.

39-GHz Band

The roll-ups of the 39-GHz band resulted in postrestructurings, with these assets mostly in the hands of Fiber Tower and IDB Spectrum Holdings Co. Each footprint license has 50 MHz of bandwidth per channel. Approximately

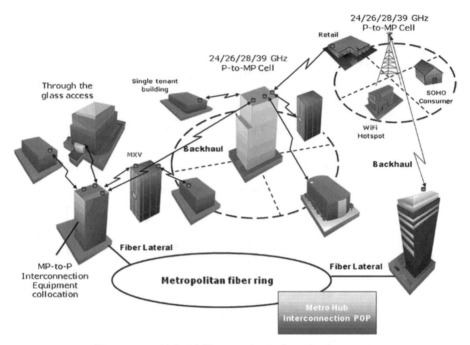

Figure 9.4. Hybrid fiber and wireless deployment.

1500 MHz of bandwidth in total is available nationwide. These frequencies were also widely licensed to competitive cellular operators throughout the world to accommodate cell site interconnection.

Multichannel Video Distribution and Data Service

There is another large swath of contiguous spectrum that is now privately controlled, but it has not yet been placed into service.

The FCC, or "Commission," created the MVDDS after an extended and often contentious process that originated in large part because of petitions from Northpoint Technology, Ltd. (Northpoint). Northpoint mounted a vigorous campaign to gain FCC approval and support for the use of the existing DBS spectrum in the 12.2- to 12.7-GHz bands.

Ultimately, the FCC determined that the service should be created, and it developed technical rules on the basis of laboratory and real-world experiments and tests to determine noninterference operations as a coprimary service with DBS operations.

The FCC issued a public notice with rules that called for an auction of the spectrum. Northpoint and its service consortium Broadwave America, Inc. (Broadwave) as well as its DBS applicant, Compass Systems, Inc. (Compass), continued to fight for an award of free spectrum and ultimately found that their litigation and lobbying efforts on Capitol Hill to gain free spectrum were failures. Northpoint and its affiliate companies elected to not participate in the auction and have no standing among any of the licensees.

The auction was conducted in two rounds, granting 214 licenses. The FCC used the Nielsen geographic descriptors' designated market area (DMA) for measuring television households, to most efficiently match the service areas to existing cable television service areas. As recently as April 30, 2007, the commission issued its dismissal of Compass's application for a hybrid DBS and complementary terrestrial network as premature and failing to recognize the MVDDS rules and auction, in which Compass and its affiliates failed to participate.

Two companies dominated the spectrum auctions, and yet neither has made any substantive deployments or established trial systems.

The FCC selected a modified version of the Neilsen DMA service area boundaries traditionally used to measure the market performance of the cable television and broadcast industries.

The MVDDS service is a fixed microwave service in the 12.2- to 12.7-GHz band authorized to provide nonbroadcast (free over-the-air services are not allowed) video and data subscription services in a downstream direction from transmitters installed in such a manner as to minimize interference with existing and future DBS receivers through the use of spatial diversity relative to the satellite transmission path. Mobile and aeronautical services are specifically prohibited in the bands.

The radio propagation characteristics of the electromagnetic spectrum in the 12.2-to 12.7-GHz bands are well known, and highly reliable predictive radio planning tools exist to support network system design. For most applications, these bands typically require line of sight between the transmitter and the receiving antenna.

The basic concept of the MVDDS frequency reuse scheme is to leverage the fixed and known positions of the DBS and the directional characteristics of the subscriber high gain satellite–receiving antennas used to receive the service. The technical rules for MVDDS define a network design that is significantly orthogonal to antennas receiving the satellite signals, sufficient to allow the satellite frequencies to be reused terrestrially without harmful interference to the DBS systems.

One of the most compelling aspects of the MVDDS bands is that future operators will be able to leverage the low cost of DBS television receivers.

Millimeter Wave 70–80–90–140-GHz Services

In 2003, the FCC adopted a Report and Order (R&O) (modified by a Memorandum Opinion and Order on reconsideration), which established service rules for the use of millimeter-wave spectrum in the 71–76, 81–86, and 92–95 GHz bands. These bands are essentially unused at the time of the R&O, and rules have been adopted to authorize high speed, point-to-point wireless links to support the interconnection of high speed local area networks, broadband Internet, and gigabit Ethernet access services. The highly directional characteristics of these beams permit multiple systems to share spectrum in proximity to one another without causing interference.

The shorter wavelengths in the 71–76, 81–86, 92–95, and 140 GHz bands permit the use of smaller antennas than what similar systems require in lower frequency bands with similar directivity and gain. The ability to leverage the huge channel bandwidths available in these extremely high frequencies (EHF) enable the delivery of fiber speed wireless services ranging from 1 gigabit Ethernet to 10 + gigabit Ethernet services.

The pioneering companies in the sub-100 GHz range include Gigabeam, BridgeWave, and E-Band Communications, each of which is selling point-to-point systems that operate at 1 gigabit Ethernet speeds.

A new breakthrough company called Asyrmatos Inc. is pioneering the first wireless solutions that can function at speeds of 10 + gigabits per second that were previously only available over fiber optic wired networks.

Third-Party Database Managers

In September 2004 the FCC released an order announcing the appointments of Comsearch, Micronet Communications, Inc., and Frequency Finder, Inc.™ as independent database managers (database manager or, collectively, database managers) responsible for the design and management of the third-party 71- to 95-GHz bands link registration system (3PDS). Since February 2005 (transition date), licensees in the 70–80–90 GHz millimeter-wave service are required to register their links through one of the database managers. The public can also access the third-party database system to obtain information about registered links. Contact information and links to the database manager's systems are at the following locations:

- Comsearch:
  http://www.comsearch.com/applications/link7090/index.jsp
- Frequency Finder, Inc.:
  http://mmradioforms.com/mmRadioForms/FrontPage.aspx
- Micronet Communications, Inc.:
  http://www.micronetcommunications.com/LinkRegistration/

10

WIDEBAND 3G TO BROADBAND 4G

Collision and Convergence of Standards

Among cellular operators, the term 4G to describe the fourth generation of mobile cellular technology has been treated as either a "dirty word" or a pariah concept that strikes deep loathing in the hearts of cellular company chief financial officers. The extraordinary financial engine that cellular services have created exists in an environment marked by incessant capital spending, with no apparent end in sight. The profit engine has been voice, and as open data-centric wireless networks emerge that allow users to freely access alternative channels for all types of voice, data, and video services. The 4G platforms, which are nearing maturity, have emerged from research and development by manufacturers and operators who have very different business models than those of the leading U.S. cellular carriers.

The following chart (Fig. 10.1) tracks the capital expense spending of both the cellular operators and the wireline telephone companies. Post the collapse of the competitive telecom market and the virtual abandonment of capital spending among the new wireline competitors, it has become apparent that the cellular operators are now routinely outspending the wireline service providers year over year. Given the capital requirements to develop nationwide-coverage wireless networks, the protectionist marketing policies that have become the norm among the US cellular carriers are logical and understandable. The development of wireless access networks has continued unabated across the globe, largely led by markets that are physically smaller and less costly to deploy nationwide services. These include the Scandinavian countries, Korea, Japan, and the United Kingdom. The

Wireless Broadband. By Vern Fotheringham and Chetan Sharma
Copyright © 2008 the Institute of Electrical and Electronics Engineering, Inc.

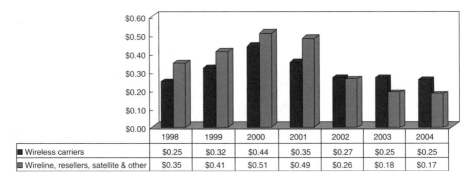

	1998	1999	2000	2001	2002	2003	2004
■ Wireless carriers	$0.25	$0.32	$0.44	$0.35	$0.27	$0.25	$0.25
▨ Wireline, resellers, satellite & other	$0.35	$0.41	$0.51	$0.49	$0.26	$0.18	$0.17

Figure 10.1.   Capex for Cellcos and Telcos.

highly populated and larger nations such as the United States, China, and India are all subject to the same large capital requirements to pay for network expansion to support basic services, much less to meet the financial pressures required to provide incremental technology upgrades (2G to 2.5G to 3G and beyond). The need to keep pace with the network capacity required to serve the rapid uptake of subscribers seeking basic services has continuously justified the cost of capital to maintain the growth of the business. The new capital spending conundrum that faces the existing cellular operators is complicated by both the massive scale of potential "forklift upgrades" to their existing infrastructure and the uncertainty of new business models coming into the market. Therefore, it is with a great deal of caution that the incumbent cellular operators in the United States are positioning their market responses to new wireless competition.

COLLISION OF 3G AND WiMAX STANDARDS

The parallel developments of the GSM cellular technical roadmap, as outlined by the 3GPP, and the continued refinement and extension of the 802.16(e) Mobile WiMAX standard will inevitably wrestle with similar technical challenges, and real-world deployment experience will benefit both camps equally.

The overlap and ultimate collision of these two standards will reflect the often-conflicting business postures of the participants. The legacy cellular infra-structure suppliers have a major vested interest in maintaining their supply contracts with service providers around the world. The extension and upgrade of the GSM infrastructure will account for billions of dollars of new business, which will fuel the continued viability of all the contributors to the value chain. On the other side of the equation, the success of the Internet-centric manufacturers include some of the most successful firms of the past decade that are extremely committed to extending their business lines into the emerging broadband IP wireless market. They have very little chance of displacing the legacy suppliers serving the cellular operators, and indeed, when Intel attempted to move into baseband chips for cellular handsets, it suffered substantial losses and exited the business. The success of WiFi and the virtual control over the destiny of the personal computing platforms has emboldened this sector of the industry to seek to lead the movement that has resulted in the rapid expansion of the WiMAX alternative for future broadband wireless systems.

How the broadband wireless infrastructure market evolves will depend greatly on which firms elect to stay purely on the 3GPP roadmap and which firms embrace both platforms that serve alternative market segments, such as Sprint, and companies that concentrate on pioneering the WiMAX platform as their exclusive solution, such as Clearwire.

The growth of the ecosystems that will support WiMAX and its reliance on open-standard, seamlessly integrated Internet networks, which are designed to be end-to-end IP from the beginning, may prove to deliver some significantly enhanced efficiencies in terms of the capital required to deploy and maintain wireless broadband systems of equivalent capacity and performance.

How the price efficiencies of the migration to pure packet networks will benefit cellular operators remains to be seen. The trade-off between the relatively high historical cost of cellular systems and the fact that the 3GPP roadmap drives through a continuum of upgrades and reinvestment in network enhancements will prevent the substantial capital savings that the green field operators will enjoy. How this disparity translates into a potential real market advantage will be tested as the early systems come online. The pure IP architecture of the WiMAX network solution eliminates the need for many of the elements required to operate a GSM network. It is too early to sample the real-world pricing strategies by the WiMAX vendors, but over the next two years we will have complete visibility into all aspects of the new generation of fully converged wireless broadband networks. It will subsequently take between 5 and 10 years for the unit volumes of WiMAX handsets and enabled devices to reach the tipping point of parity with GSM commodity pricing, but in the mean time the increased performance and flexibility will likely offset any price disadvantage the handsets face in the market. The relative positions of the 3GPP evolution and the Mobile WiMAX standard are demonstrated in Fig. 10.2 below.

Wireless broadband services exist among a number of fiber, wired, and fixed wireless services. Figure 10.3 below illustrates the relative positions of the broadband network technologies that are redefining the parameters of broadband services.

3GPP AND LONG-TERM EVOLUTION

There are a number of competing visions for the evolution of the cellular industry into a 4G packet data-centric environment.

The acronym 3GPP stands for "Third Generation Partnership Project," which is the industry trade association established in 1998. It collaborates to bridge a number of international standards organizations to develop seamless

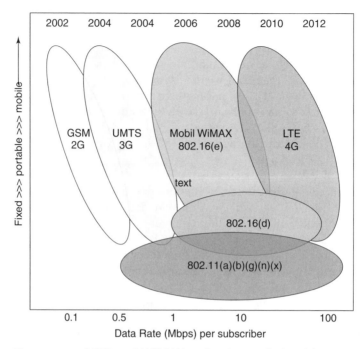

Figure 10.2. 3GPP and WiMAX technology relationship map.

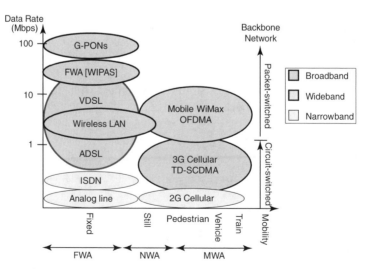

Figure 10.3. Interrelationship of broadband networks.

interoperable global specifications for the third generation of mobile systems based on an evolved GSM core network and radio access technologies they utilize (e.g., universal terrestrial radio access [UTRA], and both frequency division duplex [FDD] and time division duplex [TDD] modes). The scope of the project was expanded to also include the standards maintenance and continued development of the GSM technical specifications and reference documentation.

The 3GPP working group has developed a migration roadmap called LTE to provide an evolutionary process that will see the continued development and refinement of the GSM operator networks into 4G broadband wireless networks with pure packet-switched architectures. The LTE uplink framework is described in Fig. 10.4.

With the growing demand for increased data rates and the anticipation that cellular systems will be required to augment or displace wireline infrastructure in numerous locales throughout the world, added attention has been given to the next evolutionary steps, because the HSPA and HSPA+ standards era of increased data capacity have run their course and are unable to meet anticipated future demands for increased levels of service.

The 3GPP working group began work on the LTE standards in 2006. It is expected to deliver its final specification proposals by June 2008. However, a great deal of work has already been accomplished, and it is anticipated that early trial deployments could take place as early as 2009, although most observers do not believe that LTE will become truly relevant until much later into the next decade.

Long Term Evolution is expected to support a wide range of alternative channel sizes spanning 1.25, 2.5, 5, 10, 15, and 20 MHz. Both FDD and TDD modes are to be incorporated in the standard. The architecture is expected to use OFDMA with 2048 subcarriers of approximately 10 kHz each on the system

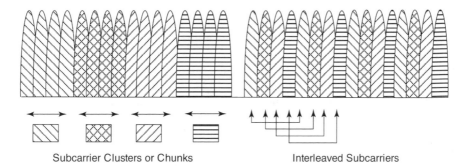

Subcarrier Clusters or Chunks   Interleaved Subcarriers

Figure 10.4. SC-FDMA uplink "chunk" channel configurations. *Source*: The 17th Annual IEEE International Symposium on Personal, Indoor, and Mobile Radio Communications (PIMRC'06), Proportional Fair Scheduling of Uplink Single-Carrier FDMA Systems, Junsung Lim, Hyung G. Myung, Kyungjin Oh, and David J. Goodman, Department of Electrical and Computer Engineering, Polytechnic University, 5 Metrotech Center, Brooklyn, N.Y.

downlink, (Fig. 10.4) and to conserve handset power, the uplink is planned to be single carrier–frequency division multiple access (SC-FDMA), which is a hybrid evolution of OFDMA subcarrier cluster configurations into selective, power-adaptive "chunks" of subcarriers, either in adjacent groups or in widely spaced interleaved subchannels. The objective of the working group is to find an optimally power-efficient method for the uplink to minimize the heavy power consumption of standard OFDMA transmissions.

A complete description of the OFDMA architecture that will be adopted in the LTE downlink is shown in Figs. 10.11 and 10.12.

The overall data rates and network capacity will vary substantially on the basis of the adaptive modulation scheme and inclusion of MIMO smart antenna technology into the LTE solution. Ultimately, LTE is a response to the recent advances made by the 802.16(e) Mobile WiMAX proponents to craft a wireless next-generation network solution that features the most logical set of trade-offs between fixed, portable, and low and high-speed mobile users. The cellular operators who entered the WiMAX arena to both learn and make contributions realized early that for optimal high-speed mobility, an FDD spectrum approach would be essential. Further, given that virtually all cellular licensees worldwide are owners of FDD spectrum allocations, the trade-offs made by the Mobile WiMAX proponents for maximum data throughput over a given amount of spectrum were deemed less important than the political and market positions that needed to be protected. Thus, we find in LTE a bias toward FDD platforms and a worthy enhancement to the incrementalist approach to refining a standard that, under the GSM banner, already serves almost 3 billion discrete mobile phone numbers worldwide.

TD-SCDMA

Time division–synchronous code division multiple access is the anointed 3G standard sponsored by the Chinese government for domestic 3G deployments and potential export. Dr. Hui Liu developed the original architecture while working at C-Will, the U.S.–Chinese partnership to develop a very high density, urban, fixed wireless telephone system. Subsequently, after the demise of C-Will, the technology was split between Navini in the United States, which pioneered its application for TDD mobile data systems, and Shin Wei in China, which pursued the original wireless telephone applications. Datang, the Chinese government–owned wireless technology consortium, initially teamed up with Siemens AG to develop the technology into a mobile cellular platform. The 3GPP has endorsed TD-SCDMA as one of the implementations of the TDD-UMTS variants of the GSM standard.

The radio access network portion of the TD-SCDMA 3G standard cellular platform functions at 1.28-Mcps (megachips per second) chip rate, as opposed to the 3.84 Mcps used in TDD-UMTS. The core network elements are shared with other 3GPP standards, resulting in some efficiencies for the equipment manufacturers to leverage the same network elements into either GSM or TD-SCDMA network applications.

Among the operators, China Mobile has said it will build a TD-SCDMA trial network when provided a 3G license by the Chinese government. Huawei, TDK, and Datang are also seeking to find export opportunities for TD-SCDMA installations and expect to find opportunities in Central and South America and Asia.

PHS and PHS II

Willcom of Japan, controlled by the Carlyle Group of the United States in partnership with Kyocera and KDDI, is the last personal handyphone system (PHS) network operator left standing in Japan post the expansion of the cellular industry into all market segments. With the consolidation, various PHS operators have merged into a single entity that is now Willcom, which has provided them with access to all of the PHS bandwidth originally licensed by the government. Functioning in the 1880- to 1930-MHz band, PHS is a TDD-TDMA microcellular architecture that uses ISDN telephone interconnection to each microcell to deliver typically 64 Kbps of data, or 128 Kbps, if two lines are bonded together for higher speed data services. Initially, each microcell was able to provide service only just about a 100–200 meters radius from the antenna site (Fig. 10.5). Subsequent smart antenna technology from Arraycom increased the coverage footprint two to three times. The original concept of low cost cellular underlay networks to be addressed by the PHS operators materialized only for a short while. As cellular prices reduced, the convenience and ubiquity of cellular proved to be sufficient enticement for PHS customers to move upmarket. To gain a better understanding of the architectural and site management requirements for a microcellular deployment, the following image (Fig. 10.6) is a depiction of the 160,000 microcells Willcom has deployed through the greater Tokyo area. The image speaks for itself.

Although PHS failed to capture much attention globally, it did manage to find a market in the second- and third-tier cities throughout China, where UTStarcom introduced the service in partnership with the local telephone companies. Ultimately UTStarcom acquired almost 80 million subscribers over PHS systems. However, these networks are now also declining in the face of the higher utility value and widespread proliferation of cellular networks (Fig. 10.7).

Personal handyphone system II is a development effort led by Willcom to leverage its established network infrastructure and customer base into the broadband future. Willcom pioneered the flat rate data plan and found a second life by delivering innovative "all you can eat" data services using the PHS network as a data underlay in advance of the widespread 3G cellular rollout and competing with DoCoMo and other 3G operators to provide low cost fixed-fee data services. Ultimately, anticipating the need to implement a broadband service, PHS II has been developed to deliver such a service, while still supporting the legacy PHS user devices and customers.

The architecture relies on the synchronous network clock capabilities of OFDMA—802.16(e) protocols that feature 5-ms time slots for transmitting and receiving data over a 5-MHz radio channel. Under PHS II, every other time slot

Figure 10.5. Microcell deployments in Tokyo. *Source*: INCREMENT P Corp.

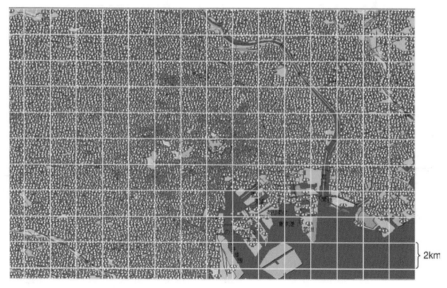

Figure 10.6. Tokyo has 160,000 microcells. *Source*: INCREMENT P Corp.

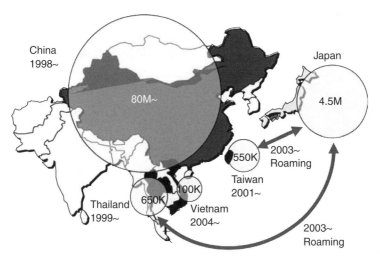

Figure 10.7. PHS subscribers by location. *Source*: Willcom.

would be allocated to the legacy PHS 200-kHz-wide channels from the legacy network, and the intervening time slots would be dedicated to broadband data services. As the customer base evolves over time and legacy PHS phones are retired, both time slots can be allocated to broadband services.

Figure 10.8 illustrates the alternating time and bandwidth slots proposed for PHS II by Willcom (Fig. 10.8).

It will take some time to realize if the Willcom initiative is successful in asserting another broadband wireless solution into the competitive landscape of nations that have allocated spectrum for PHS and who wish to evolve the technology into a next-generation alternative to WiMAX and/or 3G solutions.

WiMAX AND MOBILE WiMAX

802.16 Standard Evolution

The original working group for 802.16 was focused on the creation of a standardized common air interface for point-to-multipoint microwave systems, especially in the millimetric microwave bands above 10 GHz. All the major large-scale millimetric microwave service providers, including Winstar, Teligent, ART, and Formus went into bankruptcy in the downturn. The working group has made a lot of progress on many of the technical issues that were also applicable to frequencies below 10 GHz.

The amendment 802.16(a) evolved to address the standard's requirement for point-to-multipoint solutions in these lower frequency bands, and 802.16(d) evolved with the inclusion of TDD/OFDM into the standard and the development of the first WiMAX profiles under the standard. Subsequently the WiMAX

TDD Interleaving 300 kHz TDMA Channels and 5/10 MHz OFDMA Channels

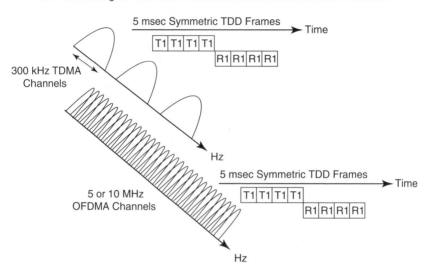

Figure 10.8.   Five-millisecond symmetric framing for PHS II. *Source*: Willcom.

Forum was organized for the commercialization of the standard and the provision of interoperability certification testing and branding control.

While the working group was concentrating on fixed applications, other developers were concentrating on the development of a mobile-capable platform that was not compatible with the 802.16(d) TDD/OFDM architecture, but that, with the addition of the OFDMA technology, would transform the standard's market potential substantially and position the working group's efforts to come into an extremely high profile conflict with the cellular standards groups.

The early developers of OFDMA included both Broadstorm, founded by Dr. Hui Liu, a professor at the University of Washington, to develop fully mobile and converged broadband wireless solutions, and Runcom of Israel, founded by Dr. Zion Haddad with an initial focus on the creation of robust video transmission systems. Figure 10.9 details the various WiMAX standards that have been developed.

Dr. Haddad worked tirelessly to get the 802.16 working group to embrace another version of the standard that was based on OFDMA technology. He ultimately prevailed, and 802.16(e) was approved by the IEEE. In parallel with his efforts on the standards committees, Dr. Liu was engaged with his team in concert with the Korean Electronics Telecommunications Research Institute (ETRI), which had searched the world for technology partners and identified Broadstorm as the most advanced in the development of OFDMA technology. The institute was working under Korean industrial policy initiatives to seek a next-generation wireless communications platform that would be exempt from the overhang of

Qualcomm patent coverage, such as the one it faced post its early embracing of CDMA technology as a cellular alternative. In addition to ETRI, POSDATA of Korea was also seeking to enter the market as a manufacturer and supplier of what became known as WiBro (Wireless Broadband), the Korean standard for OFDMA wireless broadband solutions. Ultimately, through the 802.16(e) standards process, the WiBro and Mobile WiMAX protocols converged into a common platform. During this period, which coincided with the depths of the telecommunications depression, the original Broadstorm investors decided to give up on further pursuit of their business plan and closed the company. As fate would have it, Broadstorm stopped interacting with the industry players and the standards process. At the time, this was viewed by many, including the Koreans, as the company's having gone bankrupt and the ownership of its intellectual property having thus slipped into the public domain. Unfortunately for many of the groups in Korea that possessed Broadstorm equipment at the time the company shut down, the equipment found its way into the development labs of several of the WiBro developers. Subsequently, as these organizations sought to implement a standard based on their experience with OFDMA systems, many of the unique MAC layer, channel control, and cluster management features that were part of the original Broadstorm design found their way into the 802.16(e) standard. It took Dr. Liu and his colleagues almost a year to recapitalize and relaunch the business after obtaining the assets from the former owners. The 802.16(e) working group was well advanced in its work by this time, and the new organization, Adaptix, made no recommendations to the working group for inclusion in the standard. In parallel to all these developments, the eight key core patents that covered many of the core design features of OFDMA systems were winding their way through the United States Patent and Trade Mark Office (USPTO), having never been abandoned or unsponsored by patent counsel and meeting all the requisite filing deadlines. Subsequently, more than five years after Dr. Liu filed the original eight patents covering the design of a complex OFDMA mobile communications system and largely incorporated into the 802.16(e) standard, the USPTO granted the patents. How Adaptix and the broader Mobile WiMAX community address the intellectual property rights that exist among the participants is governed by the rules of the working group, which calls for members to agree to cross-license the other members on "commercially reasonable and normal terms," none of which are specifically defined in detail. Thus, this lack of clarity may cause future challenges among the various participants in the Mobile WiMAX field. In addition, the incorporation of OFDMA technology into the 3GPP LTE standard also incorporates many of these same core design features and concepts into the LTE core architecture.

Overhyped Birth Announcements

Seldom has the author been exposed to a stranger and completely counterproductive birth announcement for a new technology platform than that orchestrated by Intel to celebrate its arrival as the chief cheerleader and credibility-granting participant to the

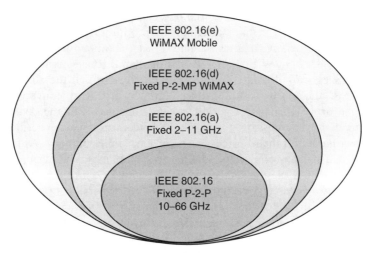

Figure 10.9. Family of IEEE 802.16 WiMAX standards.

public, the financial sector, the vendor community, and the potential customers for the infrastructure equipment. While at a conference in Australia, the author was exposed to the "script" that was being read all over the world at almost the same time. Whoever had drafted the announcement of their participation in WiMAX seemed to have picked one truth from virtually every application and domain that would be touched by this new technology. Unfortunately, none of the claims were ever likely to exist at the same time in the same application. Therefore, the global press was suddenly touting the Intel miracle, announcing that our laptop computers would soon be able to communicate over distances of 15 miles at data rates in the 60-Mbps range. Yes, our laptops will inevitably be WiMAX equipped, and they will communicate with cellular-like base stations, but more likely on a footprint quite similar to legacy cellular systems between 0.5 and 2 miles. In many cases, WiMAX networks will function over microcells or picocells in which the range typically varies from hundreds of feet to 500 m.

As for the long-range transmission distances, those claims are based on point-to-point long-haul applications using large directional high gain antennas. The data rates will surely reach these levels, but the primary uses will inevitably be over relatively short ranges (Fig. 10.10).

Mobile WiMAX: 802.16(e)

The ability to provide fully converged broadband wireless services to low cost devices requires a fresh look at how our wireless data services are delivered. The rapid shift in attention by virtually all of the WiMAX Forum major participants from 802.16(d) to the completely non-backward-compatible 802.16(e) OFDMA-based solution is one of the recent marvels of the standards creation process. The clear superiority of the adaptive dynamism in all the domains of a radio system

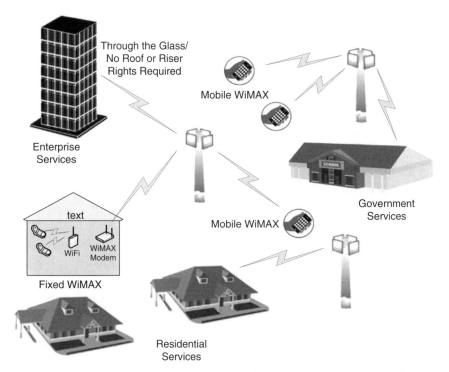

Figure 10.10. A graphical representation of the various WiMAX topologies.

inherent in TDD/OFDMA over the capabilities of the fixed TDD/OFDM "flat channel" platforms made it relatively easy for a number of participants in the WiMAX Forum to quickly embrace the fact that the mobile solution could do everything the fixed solution could do, proving better in every respect, especially in terms of spectrum efficiency.

Why Will OFDMA Prevail in Broadband Wireless?

A number of unique capabilities are inherent in OFDMA wireless technology. Specifically, it is the first wireless system level solution that has the ability to dynamically control every domain of the radio system. There is substantial benefit in terms of spectral efficiency with OFDMA, which in an optimized converged network deployment can enjoy over a fourfold increase in capacity, using a given amount of radio spectrum, compared with CDMA platforms. In less optimized environments, there is at least a two times increase in capacity over CDMA solutions at the same power and frequency. Thus, the increased demand on carriers to provide services to an ever-growing number of subscribers at constantly increasing data rates, while constrained by the limited amount of mobility optimal spectrum below 3 GHz, will ultimately compel the industry to migrate to OFDMA-enabled efficiencies.

OFDMA = OFDM + FDMA + TDMA + CODING/TDD

- Has the multipath resistance of OFDM
- Has the flexibility of TDMA
- Has the adaptability of FDMA
- Has the spectral efficiency of TDD
- Has the benefits of coding
- Also leverages intelligent resource allocation
- Has control over the time domain
- Has control over the frequency domain
- Has dynamic frequency selection—assigns groups of subchannels based on signal to interference and noise ratio (SINR), the QoS required by the application, cochannel interference, and self-healing reconnection
- Has dynamic power control—adjusts the power output of the subscriber device to maximize its modulation rate and to minimize its interference with other subscribers
- Has dynamic time slot allocation—assigns time slots for each subscriber according to provisioned class of service, available subcarriers, session data load, QoS bandwidth reservation, and priority queuing

Constant Striving for Maximum Efficiency

The OFDMA system is in a constant state of adaptation, a dynamic state always seeking to provide every communication at the highest level of modulation using the least amount of spectrum to deliver the quality of service required by each individual data transfer session, all managed in real time.

"Orthogonal" as Applied to an OFDMA System

Orthogonal frequency division multiplexing is a very special form of frequency division multiplexing (FDM), where the carrier frequencies are spaced much closer together (to the extent that subcarrier guard bands disappear) to create many more carriers within a given bandwidth.

- The large number of subcarriers enables the information-carrying capacity of a given amount of spectrum to be significantly increased.
- The more the usable subcarriers, the greater the capacity of the available bandwidth.
- Normally, modulated carrier frequencies spaced too closely together would interfere with each other, as shown on the diagram below (Figure 10.11).

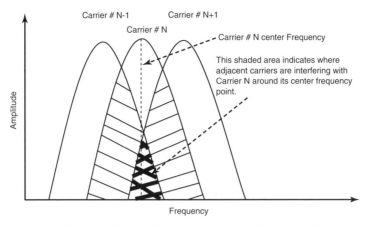

Figure 10.11. Interference from adjacent modulated carriers. Orthogonal condition achieved by eliminating the heavy shaded area.

The challenge, therefore, is to make these carriers noninterfering or "orthogonal," which is achieved by a combination of techniques such as (1) special selection of the number, frequencies, and the spacing of carriers and (2) the intelligent use of phase shift keying (PSK) and quadrature amplitude modulation (QAM) techniques.

Frequency Spacing of Carriers

- Carriers spaced very closely together can be made orthogonal (not causing mutual interference) provided the frequency separating each is carefully chosen.
- In OFDM, a carrier is modulated using PSK and QAM techniques for a certain period of time called a symbol period (T).
- Therefore, the periodicity of symbols is $1/T$ in Hz, i.e., $F = 1/T$.
- The frequency spacing between carriers mentioned above is carefully chosen to be an integer multiple of F (i.e., $1F$, $2F$, or $3F$).
- If these conditions are met, the potentially interfering $N-1$ and $N+1$ carriers will exhibit a null at the center frequency of carrier N and will be mutually orthogonal. In a multicarrier system with multiple side bands from all carriers, the desirable condition would be achieved at the center frequency of every subcarrier. Note this can only be achieved in relation to interference from all subcarriers in the same cell. Therefore, careful RF planning is still required for multiple cell deployment.
- If the receiving filters that define the spectrum over which the demodulator operates are designed to be very sharp, the interference that does exist will not exceed the acceptable signal-to-noise ratio (SNR) level. Note that the

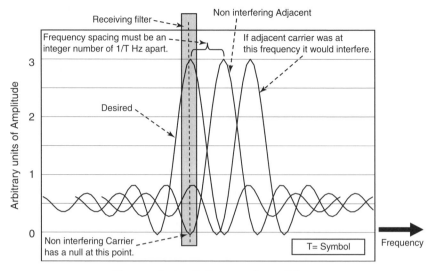

Figure 10.12. Orthogonal condition.

demodulator at the far end needs only a very narrow bandwidth around the center frequency of each subcarrier to detect the phase or amplitude modulation change. The interference, which is present just outside this band, is therefore not detected.

- This orthogonal condition is shown in the above image (Fig. 10.12).

Subcarriers

- In theory it would be possible to have a thousands of separate subcarriers spread across a particular amount of available bandwidth, but ensuring that the orthogonal is maintained within a cell becomes increasingly difficult.
- Therefore, until recently in practice the number of subcarriers used varied between 64 (WiFi) and 256–2048 (WiMAX range of all versions).
- With the very basic form of OFDM, all subcarriers will be used to carry different parts of a single user's message, i.e., at any point in time, all usable subcarriers carry the traffic from a single customer.
- In practice, in a working system, some of the subcarriers will not be used because of poor SNR or for creating part of the guard band.
- Having established a number of orthogonal subcarriers, we can now modulate the "content signal" onto all of the usable subcarriers using PSK or QAM techniques.

Carrier Clusters

OFDMA implementations are constantly sampling the availability and performance of each subcluster of subcarriers that have been selected for a specific real-time payload assignment, singly or in as many groups as are required to meet the tasking.

Figure 10.13 is a screen print of an actual real-world trial of an OFDMA system in a high density urban environment. Each of the 32 separate constellations shown represents a dynamic assignment of a specific modulation scheme to specific clusters of 16 individual subcarriers. Heavily faded regions of the available 5 MHz of bandwidth can be seen in the clusters that cannot realize a tight constellation. The power of OFDMA is its ability to tolerate such fades and nulls in real time and allocate the remaining bandwidth at the highest order of modulation to deliver the largest amount of data throughput through a given amount of available radio channel capacity. This inherent interference mitigation and tolerance capability on a continuous basis is one of the key aspects in delivering spectrum efficiency in OFDMA systems.

Fully integrated OFDMA-based mobile broadband IP networks deliver fully converged network services.

- Video
- Data
- Voice

The OFDMA technology also had the advantage of simultaneously delivering all ranges of data speeds within a single radio spectrum channel with no loss of efficiency.

- Narrowband
- Wideband
- Broadband

Also of significant merit is that OFDMA systems provide the best compromise to support fixed and mobile convergence within the same network. Thus, the same network can simultaneously support highly efficient service delivery to all user locations and domains.

- Fixed
- Portable
- Mobile

The resultant business benefits to service providers and carriers include

- lower capital expense/subscriber
- lower operating expense/subscriber
- higher data throughput/subscriber

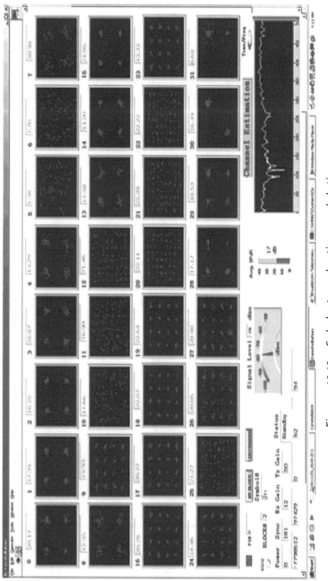

Figure 10.13. Subcluster adaptive modulation.

- greater number of serviceable subscribers/sector and given amount of radio spectrum
- greater flexibility for spectrum deployment

It is anticipated that the parallel convergence of the GSM cellular LTE (primarily used in FDD networks) and the Mobile WiMAX (primarily used in TDD networks) standards will result in a wide range of network topologies ranging from traditional macro coverage cellular configurations, microcellular deployments, and underlay network extensions, including picocells and femptocells, with a plethora of backhaul and interconnection network alternatives ranging from fiber to wireless (Fig. 10.14).

Mobile WiMAX Versus GSM LTE

There is a continuum of development in our industry, which leverages the regular advances in the applied sciences of materials, computing, software, systems, power storage and creation, and network topologies and routing algorithms.

When industries as large and successful as mobile communications are faced with new forms of competition, the suppliers and implementers of the incumbent ecosystem inevitably must see the world and its future through the lens of their

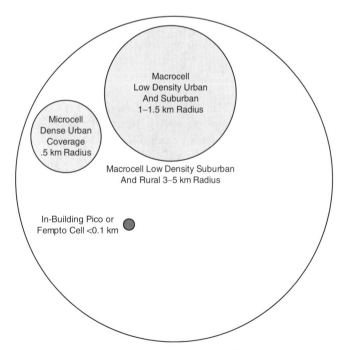

Figure 10.14. Stratified wireless broadband networks.

own success. Therein is the central theme that defines the intersection of the alternative paths to a 4G wireless world. One path being planned is an incremental extension of the circuit-switched (CS) legacy core network into a hybrid CS and packet network, all aiming for a pure packet network future. Such an approach makes every sense for incumbent operators who are still digesting billions of dollars of sunk capital in their existing networks. The other path that seeks to arrive at virtually the same 4G future is being driven largely by the computer and Internet sectors, which have evolved to a point where they are convinced that the traditional cellular networks do not have the unfettered growth and open network concepts that reflect the full potential of the Internet. This conviction has been reinforced by the enormous success of WiFi in advancing the wireless local area network into a global wireless phenomenon that has seen this simple technology spread into service niches and applications far beyond its original design purpose. Although the WiMAX Forum membership includes a number of legacy cellular carriers, its agenda is still largely being set by the wireless Internet and computing camps; the legacy cellular world is largely following the lead of its core suppliers in seeking to keep "forklift" upgrades to a minimum.

The primary variance between the two camps seems to devolve into one primary difference in managing the radio spectrum resource. Simply put, cellular mobile systems thrive in FDD environments, and Mobile WiMAX networks are designed to simultaneously support a mixture of all types of services, including fixed, portable, and mobile, simultaneously. The WiMAX proponents have selected TDD, using a single radio channel to manage both uplink and downlink data flows. This approach has been selected to deliver the most efficient performance in a fully converged data-centric service environment (Figs. 10.15 and 10.16).

FDD Advantages

Mobile system benefits from FDD include the following: (1) The cellular industry was founded on one underlying principal—mobility. The use of separate frequencies to transmit and receive signals is essential to deliver wireless services to high speed mobile devices because of all the channel conditions that would degrade the signal, such as (a) fast-fading environments (moving behind vehicles or buildings

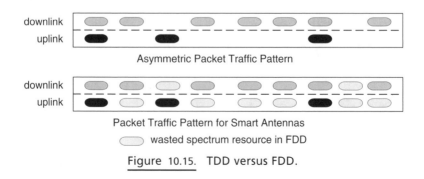

Figure 10.15. TDD versus FDD.

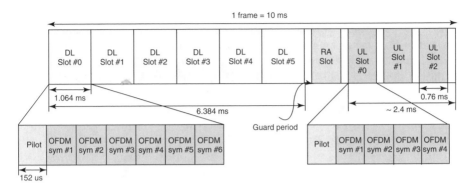

Figure 10.16. Guard-time separation for uplink and downlink.

that block the radio signal), (b) Doppler shift (varying speeds constantly changing the frequency stability of the signal), and (c) guard bands separating the downlink from the uplink spectrum (provides large separations between the transmit and receive channels, thus preventing adjacent channel interference caused by exposure of low power user devices to high power base stations).

TDD Efficiency Advantages

Converged data-centric systems benefits from TDD are as follows. (1) As shown in Fig. 10.15, TDD is able to pack more information packets into a given amount of radio spectrum. Because most data traffic is asymmetric, using the same radio channel for both the upstream and downstream transmissions of data ensures much less wasted space compared with FDD systems. (2) Time division duplex systems also have the advantage of two less guard bands protecting the edge of the channel(s) compared with the FDD systems. (3) The TDD system uses time-guard spaces to separate the data traffic flowing up- and downstream. There have been substantial improvements in the efficiency of managing time "guard bands" over the past decade, and time (capacity of slots) overheads have reduced to less than 5% even when consideration is given to large radii out to 10 km. This is a substantial improvement over the frequency guard bands required by FDD systems, which on average will waste about 15% of the available bandwidth as null zones along the two extra channel edges required in full duplex channels (double that required for TDD systems.)

Thus, both FDD- and TDD-based systems will enjoy their own unique attributes that are key to delivering the services defined by their historical market position. Mobile WiMAX is not a replacement for high quality robust FDD-based mobile systems. It is however, a highly efficient means of delivering a fully converged suite of quad-play services that extend out of the mass-market experience with the Internet.

With this distinction stated clearly, we hope that the industry debate on which technology is "best" can abate, and technologists and product managers can concentrate on developing services and feature sets that are optimized to meet the needs of their installed base or future broadband customers. There is little chance that either the 3GPP or Mobile WiMAX camps will have unilateral control over the final configuration of the 4G wireless broadband platform. However, there are large areas of technology overlaps that will contribute to the ultimate outcome of these parallel technology roadmaps and evolving ecosystems, regardless of their ultimate convergence or continued parallel evolution.

International Regulatory Parity for Mobile WiMAX

On October 19, 2007, the ITU adopted Mobile WiMAX under the 802.16(e) standard as one of the approved protocols for international wireless mobile systems.

"Including WiMAX in the IMT (International Mobile Telecommunications) category sets the stage for the WRC-07 (World Radio Conference) to consider the future spectrum availability for next generation broadband mobile services," according to Ambassador Richard Russell, associate director of the White House Office of Science and Technology Policy and head of the U.S. delegation to the 2007 WRC.

The endorsement by the ITU, which was made just in advance of the WRC-07, finally levels the playing field between WiMAX and the dominant existing cellular standards under the GSM and CDMA protocols. The previous lack of ITU endorsement was an obstacle for numerous WiMAX proponents around the world who faced regulatory uncertainty from their local regulators because of the lack of guidance and endorsement for the standard from the ITU. National regulators will now be much more likely to allocate spectrum for use in the TDD deployments preferred by the leading Mobile WiMAX advocates. Further, in some countries, there have been specific prohibitions on the deployment of non-ITU-approved technologies under the IMT standards.

Mobile WiMAX OFDMA TDD will now take its place alongside the only five other existing approved global mobile wireless standards listed below.

- CDMA2000—the Qualcomm-based platform leveraging 1.25 MHz wide modular channels.
- W-CDMA—the 3GPP GSM standard for broadband wireless that uses 5-MHz FDD channels.
- UWC-136—the broadband wireless standard developed by the Universal Wireless Communications Consortium. UWC-136 was developed in the United States as an evolutionary path for the original analog Advanced Mobile Phone System (AMPS) and the 2G TIA/EIA-136 TDMA standard.
- DECT—Digital Enhanced Cordless Telecommunications is an ETSI standard originally developed for digital portable (cordless) phones. It is also recognized by the ITU as a 3G system. IMT-2000 has labeled DECT as

IMT-FT (frequency time). DECT operates in the 1900-MHz and Unlicensed Personal Communications Services (UPCS) frequency bands.

- TD-SCDMA—Time division–synchronous code division multiple access is the 3G standard developed under the sponsorship of the People's Republic of China by the Chinese Academy of Telecommunications Technology (CATT), Datang, and Siemens AG. The standard was included in 3GPP Rel-4 and is known as the "UTRA TDD 1.28-Mcps Option."

11

RADIO TECHNOLOGY

Moving the Goal Posts

ENABLING TECHNOLOGIES

Wireless infrastructure is shrinking in terms of size, complexity, and cost. This statement does not however mean that it is getting less sophisticated or technically challenging. As the power of microprocessors has increased, wireless devices have gained in terms of performance, flexibility, applications, and speed. The next generation of wireless networks will benefit from a number of enhancements to the underlying technologies that contribute to their system performance.

The areas presently holding great promise for enhancing our wireless broadband systems include

- materials sciences
- digital signal processing
- modulation enhancements
- antenna technologies
- reduced power consumption microprocessors
- reduced thermal heating of microprocessors
- battery and power supply improvements
- display technologies to enable larger form factors in small packages
- plug and play user self-installation and network provisioning
- the overlap of photonic and radio system technologies

Wireless Broadband. By Vern Fotheringham and Chetan Sharma
Copyright © 2008 the Institute of Electrical and Electronics Engineering, Inc.

Materials Science

The applied sciences continue to enhance the performance of key components required to develop advanced wireless systems. Listed below are key areas that are presently capturing significant attention of the suppliers and developers of radio system components to address shortfalls in current technology.

Solid State Power Amplifiers. One of the most important components in a broadband wireless radio system is the power amplifier, and its inherent stability and ability to meet rigid emissions mask cutoffs to enable tight intercarrier spacing between adjacent RF channels.

RF Synthesizers. Our industry is still anxiously waiting for reasonably priced synthesizers capable of tunable bandwidth ranges greater than 500 MHz. The major missing component for software defined radios (SDRs) is the ability to tune among discrete frequencies over the entire range of protocols that add value to a flexible, dynamic device. For example, the ability to synthesize any frequency from 700 MHz to 3 GHz would allow a device maker to make SDRs that could morph into virtually any of the mobile communications standards.

Dynamic Range Digital Filters. The ability to isolate radio frequencies within a single device that operates in multiple bands is limited currently by the lack of dynamic range of traditional RF filters.

CMOS Enhancements to Replace Reliance on GaAs MMICs. Gallium arsenide (GaAs) has been the material of choice for highly stable RF devices above 2 GHz. It is still extremely expensive to use in price-sensitive high volume products. Recent enhancements to the low cost, low power-consuming, and easier to manufacture CMOS chip technology are enabling the replacement of GaAs components, which are extremely expensive, with improved performance CMOS and CMOS/GaAs hybrid chips.

The continued reduction in the size of transistors and a corresponding increase in density are enabling CMOS and silicon-germanium (SiGe) to replace GaAs in microwave and low power consumption applications such as mobile wireless devices, personal computers, GPS receivers, and satellite digital set-top boxes.

Lateral Double-Diffused MOSFET. LDMOS has demonstrated high performance capabilities that were only possible with GaAs devices a few years ago. As frequencies have increased and linearity and efficiency requirements become essential to developing broadband wireless systems, LDMOS components with output power over 100 W at 2+ GHz have emerged to deliver the increased performance required in a silicon-based technology, keeping costs reasonable. Applications to WiMAX broadband wireless base stations that require high linearity requirements at frequencies up to 3.5 GHz are presently in process.

Digital Signal Processing

The ability to implement SDR and cognitive radio technology is dependent on the availability of DSPs with sufficient power to be able to manage all of the computational requirements needed to emulate all of the waveform creation, modulation, and coding/decoding required to implement a wireless system. Digital signal processors have been almost universally selected by the OEM manufacturers of WiMAX base station equipment and are also being widely used by the device manufacturers for their prototypes, which cannot wait for the system-on-a-chip (SOC) solutions to arrive before testing and modeling their solutions or for low volume specialty products that will not justify the expense of building a dedicated ASIC.

The technical challenges with DSP-based solutions touch on four key aspects of radio system performance.

1. The amount of raw processing power that is available.
2. The cost of even the most powerful DSP constitutes only a very small percentage of the finished base station products. With mobile handsets, however, price sensitivity is extreme, and the cost of the chip is a crucial element that must be considered to achieve success.
3. Power consumption is a challenge for DSP-enabled battery powered devices, as they consume substantially more power than ASIC chip.
4. Heat generation is another by-product of powerful DSPs, which is easily managed in base station scale infrastructure, but which presents significant challenges in user terminals.

Modulation Enhancements

The incorporation of adaptive modulation into the Mobile WiMAX standard is a key contributor to the overall system efficiency.

Future enhancements to the standard for higher order modulation options to be incorporated into short-range picocell environments will enhance the overall flexibility of Mobile WiMAX to match up to the future needs of users for data rates in excess of 100 Mbps. The hybridization of the architecture to support all of the common domains that Mobile WiMAX networks will need to address the following:

- High speed vehicular mobility—maintain >2 Mbps at highway speeds
- Pedestrian mobility—experience useful data rates of >10 Mbps
- Portable and fixed devices in all locations—spoof 100 bT (metro Ethernet)
- Fixed mobile convergence—spoof gigabit Ethernet 1000 bT at data rates >100 Mbps
    - Enterprise—extend the IP PBX and corporate LAN via VPN
    - Home —leverage the terrestrial broadband connections in homes, whether cable modem or DSL based, to maintain all features and functions of the network capabilities (either 3GPP or WiMAX) in the home environment.

○ High speed WLAN—provide the benefits of VPN and submask virtual local area network (VLAN) capabilities to address the ability to have access to home or office files and applications, regardless of location.

The other adaptive domain facing enhancement requirements is in the partitioning of the TDD upstream and downstream framing balance. The ability to have the boundary between the upstream and downstream data flows will be an important capability of future wireless broadband networks. The ability of the wireless system to sample its traffic in real time will increase the utility value on specific spectrum blocks.

The traditional mobile cellular operators have also been positioning for an FDD version of OFDMA termed Long Term Evolution (LTE) to allow for its seamless integration within targeted cellular markets. This alternative will provide the well-proven and standard FDD approach traditionally used in mobile cellular networks, less efficient from an ultimate data throughput capability, but very robust for high speed mobile communications.

Antenna Technologies

Mobile WiMAX has incorporated the smart antenna technology into the standard. How the SIMO (single input, multiple output) and MIMO (multiple input multiple output) techniques are adopted by service providers for integration into the network will become obvious over the coming 24 months.

Additional antenna investigation should concentrate on the indoor require-ment for high gain link margin-enhancing capabilities. The potential for tracking antennas, dynamic beam-forming antennas, and low cost printed antennas embed-ded in a wide range of devices will be worthy areas of investigation by the engi-neering and product development communities.

Reduced Power Consumption Microprocessors

The industry's efforts to increase the battery life and useful performance duration of mobile devices are being led by the device and chip manufacturers, as their improvement curve is proving to be more rapid than that enjoyed by the battery industry.

Reduced Thermal Heating of Microprocessors

As the computational power of microprocessors increases, there is a corresponding increase in the heat generated by the devices. The field of heat management holds substantial promise of improving the user experience of future mobile devices.

Battery and Power Supply Improvements

There is a global initiative to improve the performance of batteries for use in mobile devices. There is research in basic materials, new form factors, and even micro fuel cells. In addition, portable charging devices are enhanced to reduce their weight and form factors and increase the flexibility of connectors to support a wide range of plugs and electrical power utility voltages and amperages.

Display Technologies to Enable Larger Form Factors in Small Packages

There is ongoing development for new generations of flexible polymer display technologies that will enable larger screens to be stored in such a way as to allow them to be unrolled or unfolded to provide a much larger user display on small form factor mobile terminals.

Plug and Play User Self-Installation and Network Provisioning

One of the areas that can deliver substantial enhancements to the efficiency and profitability of service providers of all types is the creation of operational support systems and application software that enables user devices to have "self-discovery" and "self-provisioning" capabilities on broadband IP networks. The time and expense of customer support staff to provision each new device onto a network is substantial, and the improvement in efficiency will dramatically improve the customer's service experience.

GENERIC RADIO DEVICES

There is a growing bifurcation in the development of radio system infrastructure and user devices. The traditional path is along a hardware-specific trajectory, and the other is with fully software-defined, completely generic devices, similar in concept to the personal computer, which is just a platform until it is given software instructions regarding the specific task required. The mainstream approach that uses discrete components, with specific tasks and functions, is hardwired into firmware and devices dedicated to delivering specific services on predetermined network protocols and at assigned frequencies. The path that this hardware-defined portion of the industry is pursuing has seen the integration of multiple protocols and multiple frequency bands into single devices that have the ability to function on different systems, but only a few preloaded protocols are practically following this approach. Today we have handsets in the market that can roam onto cellular networks operating on 800, 900, 1800, and 2400 MHz, using discrete radio components to support each of the different frequencies and embedded software to support each of the separate operating systems and protocols ranging from analog to TDMA digital to CDMA, but each application is built into the device as a distinct, individual subset, all tied to firmware and integrated into a multiprotocol single device.

The other branch of device development that is gaining momentum is SDRs, which morph into whatever type of service platform their software programing instructs. The underlying hardware comprises DSPs and logic chips with sufficient power to control all aspects of the radio channel, its modulation, and the encoding of information into whatever service descriptors are required, be it voice, data, video, or audio information. Software-defined radios are the core foundation of the next wave of flexible and adaptable radio system devices under the broad heading of cognitive radios.

Software-Defined Radio

SDR technologies will bring dynamic protocol and frequency adaptability to devices that will be under the control of the user. Software-defined radio devices are essentially computer-controlled radios that create all of the waveform generation, channel configuration, and modulation scheme implemented in software instructions for mixed signal (analog and digital processes on a single chip) DSP platforms. Early implementations of SDR radios have been proliferating among the WiMAX manufacturers for use in WiMAX base stations and terminals that are coming to market before the interoperability standards testing has been completed by the WiMAX Forum. These SDR-based radio devices have the ability to be reprogramed remotely through software downloads when the standards are finalized. The authors have been deeply involved with the early development of Mobile WiMAX equipment created using SDR-controlled systems, including base stations and user terminal devices. The performance of these software-controlled devices was virtually indistinguishable from similar systems that were implemented using traditional discrete radio devices.

The largest challenges facing developers of SDR-based systems include (1) the cost per MIP of mixed signal DSP chips, (2) the relatively high power consumption of the DSPs and the related heat generation, and (3) the present limitations of frequency synthesizers to tune dynamically across multiple gigahertz of bandwidth. As these difficulties are resolved, and there is presently a massive wave of development in the research laboratories around the world to solve these issues, the creation of practical and powerful generic platforms will join the quest for commercial success.

Software-Defined Radio: Physical Layer

The RF synthesizer technology and DSP processing power together determine the frequency agility of terminal devices, which determines how wide a portion of the spectrum that the terminal device can scan looking for an access network base station signal. Military equipment for high value national security applications can scan and operate over several gigahertz of spectrum (small unit operation situations awareness system [SUO SAS] from DARPA/ITT Aerospace and Communications scans more than 2 GHz). Digital signal processor products from Sandbridge and PicoChip are promising newcomers that could provide the processor speeds required. To date, commercial products vendors have provided cost-effective multiband solutions based on multichip implementations that have $\pm 10\%$ frequency agility for the synthesizer. Next-generation commercial synthesizers are expected to be available to

- reduce the number of product versions required to serve a global market-place where mobile broadband wireless access frequency allocations are unmatched
- allow operators with frequency allocations in small and/or less frequently allocated bands within this range to participate in the global roaming opportunity

This is very good news for operators, regulators, and the general public, as more services over more networks and users are potentially accessed with the same basic equipment. The marketability of such terminal devices increases as coverage becomes universal, and costs come down as the component volumes increase.

Future users of next-generation protocol and frequency-agile devices will be able to choose among a plethora of underlying access and/or service providers in real time. The power to control their access choices and have discrete control over the selection of content from virtually any source is a vision that is coming into focus. The devices that will empower this dynamic flexibility will evolve out of the present research into SDR components and the growing intelligence of cognitive radio control systems.

Cognitive Radio

Cognitive radio technology brings a higher level of artificial intelligence to the sphere of radio devices. Cognitive radios essentially take control of the "knobs and dials"* of the SDR platform, enabling artificial intelligence to sense the environment and define the assignment of frequency and protocol to meet a specific objective, all in real time and on the fly. Users will provide instructions that will populate "rules tables" that the cognitive radio system will apply to engage with the changing radio situational landscape and service provider environments. As users move between various networks and seek access to various types and kinds of services, these cognitive devices will morph into whatever RF device is desired by the user. Future users of these technologies will be able to select services on demand or as they move between coverage patterns of various networks.

A typical scenario may find users cycling their device among multiple networks: (1) at one moment making a cellular call over a subscription network; (2) then, when finished, a quick review of the digital television stations in range, nothing appealing; (3) switch to the Internet over a 700-MHz Mobile WiMAX system to review the progress of the search bots previously assigned to a personal research project; (4) search for WiFi access points that allow access to your favorite VoIP service; (5) on finishing your VoIP call, you decide to tune to your music subscription on Apple Radio, where you decide to purchase some

---

* Dr. Bostian, Virginia Tech, Blacksburg, Virginia.

downloads that require you to open an Internet connection with any available wireless resource in your range. All of these typical profiles could have been tuned and authenticated automatically by the cognitive radio on the basis of real-time instructions either from the user or automatically via preselected profiles determined by the user that could reflect time, location, or context of various services.

Digital Fingerprints

The use of ESNs to identify specific wireless devices for authentication onto wireless networks has been the most common method of controlling unauthorized access by devices onto shard public networks. Over the years, a wide range of fraudulent attacks have been launched against operators by illegal means, which were often based on cloning legitimate ESNs for illicit use, masking the identity of the illegitimate user through the use of the spoofed device identity. Currently, promising research is under way at Virginia Tech* to develop the ability to sample the digital signature of specific wireless devices that do not require any ESN or embedded identification system to enable a network operator to identify and certify these as legitimate user devices with appropriate authorizations for access to services that make use of only the network's ability to sample the "digital emissions signature" of each and every unique device on the network in real time.

This technology holds the promise of dramatically reducing carrier exposure to fraudulent-use revenue losses. It is also likely to improve and simplify the device authentication process and the network overhead associated with dedicated control channels (or control time slots), while reducing the complexity of core network databases that have been developed by the wireless industry to control roaming and intercarrier access control. It will be several years before this electronic signature technology likely finds its way into commercial service. We bring it forward in this work as a typical and promising example of the continuing stream of refinements and developments that the engineering community continues to bring into the wireless broadband environment.

---

* This research is being conducted by Kyon Woung Kim, graduate research assistant, Virginia Tech, Blacksburg, Virginia.

12

CONTENTION AND CONFLICT

Regulatory, Political, Financial, and Standards Battles

This chapter provides some background and insight into the regulatory, political, and financial conflicts and issues that need to be addressed. As we enter the next wave of competitive engagement between the well-established and powerful incumbent service providers and the emerging convergence players, including the expansive software industry, mature cable system, and satellite broadcasting operators, it is going to be extremely interesting to see how new types and kinds of broadband wireless service providers will fare against the assembled power and protectionism afforded to the legacy providers by regulators and the financial community.

As the future unfolds, the influence of the various elements participating as contenders for leadership and sustainable market share in the wireless broadband future will battle continuously to influence the regulatory, political, technical standards, and financial decision makers regarding why their unique situation is worthy of defending or enabling. As we have discussed previously, there is limited incentive for mature incumbent operators to innovate or invest in network facilities and applications that either cannibalize existing profitable services or require the deployment of high cost infrastructure that displaces existing functional equivalents. Thus, we are constantly searching and hoping for the breakthrough technology or product that will spring an iPhone or iPod-like surprise on the real-time communications market.

Wireless Broadband. By Vern Fotheringham and Chetan Sharma
Copyright © 2008 the Institute of Electrical and Electronics Engineering, Inc.

REGULATORY DRIVERS

The large unfinished job across the world of opening national markets to global competition and participation by multinational service providers is often tagged "market liberalization." The introduction of competition into the telecommunications market was widely assumed as a logical regulatory strategy to drive prices down in the marketplace and to multiply the services offered to the public. However, it has been an ongoing and lumpy process since the breakup of the U.S. monopoly in 1984 began the global process. In the European Economic Area, 1998 was the big year for liberalization, and the liberalization of the communications sector in developing countries is the next major wave that is well under way.

Moving from monopoly to competition in the communications sector was supposed to begin an evolution from expensive, innovation-resistant, limited, or no-choice-for-alternative services to an environment encompassing lower cost, innovative services, and a wider range of choice among competing service providers. Where cellular wireless services have been the primary new competition to the ILECs, there has been a continuous drive toward full and empowered competition on autonomous competitive networks. On the wired side of the market, only the fiber and coaxial hybrid networks of the cable system operators have made much progress with the development of services that are legitimate alternatives to the copper telephone networks. Virtually all of the CLECs exist to skim resale arbitrage from the legacy access services of the ILECs. As the market matures for NGN services leveraging metropolitan broadband IP networks owned by new entrants, it will mark the first time that the CLEC competitors will have a legitimate opportunity to deliver services that are largely still unavailable from the ILECs.

Ironically, however, in the United States, astounding market and financial power has been reconsolidated under the control of the two leading ILECs, which led the charge to reconstruct the former Bell monopoly position of dominance, namely Southwestern Bell, now AT&T, and Bell Atlantic, now Verizon. In the authors' opinion, the federal and state regulators who had approved the series of mergers and acquisitions that resulted in the new AT&T and Verizon will be proven to be shortsighted at best, and their failure will likely be considered as a dereliction of duty by future generations. Their failure to appreciate, much less honor, the original objectives of the consent decree issued by Federal Judge Harold Greene in 1982 between AT&T and the Federal Trade Commission (the MFJ) has resulted in the creation of organizations with market and financial power, which have the power to literally crush virtually any of the existing or putative players that encroach upon their newly created and largely unfettered hegemony in the converged telecommunications services market. Bush administration regulators have made the preposterous assumption that true competition had matured within the telecommunications industry, demonstrating false confidence that the cable operators, CLECs, ISPs, and independent Cellcos were somehow sufficient to offset the power of accumulated incumbency represented in the new dominant players. Previously restricted geographic ILEC market locations have dissolved in the face of these two companies' control over nationwide cellular networks, their ability to provide long-distance services, and ownership of a substantial portion of the national Internet backbone network infrastructure. Putting paid to this observation is the recent outcome of the FCC's 700-MHz auction in which AT&T and Verizon together accounted for almost 90% of the winning bids, thus ensuring that no new national mobile broadband wireless operator will emerge under the present U.S. telecommunications public policies. A refreshing counterpoint to this outcome is the Canadian government's approach to its AWS auction in May 2008, which mandated that 40% of all spectra on offer had to go to companies that were not presently national wireless operators. This approach will ensure that new competition will emerge in Canada to challenge the existing "tri-opoly" that presently exists among Rogers, Bell, and Telus. The United States should benefit from the adoption of such policies as a viable means of creating legitimate, empowered competition to the current virtual "du-opoly" environment comprising Verizon and AT&T that has emerged to dominate the U.S. market.

Further, regulatory relief over mandated resale of new digital facilities such as fiber-optic and DSL upgrades to the analog copper access network has destroyed almost entirely the competitive inducements that were inherent in the Telecommunications Act of 1996.* The protection and encouragement of the newest competitors was established as a fundamental tenant in the act. However, over the past decade we have witnessed the real world and suboptimal outcome of the public policy agenda mandated by Congress for the creation of true competition. Unfortunately, the act, which mandated various controls and obligations on the ILECs to both empower a "level playing field" between the new competitors and the incumbents by placing price caps on copper access network circuits, this approach has resulted in a resale market that exists primarily upon arbitrage. Significantly, this heavy reliance on the resold copper access networks has resulted in a landscape wherein there were few incentives and scarce capital resources for new players to construct their own new access facilities, which could be used to provide higher bandwidth access services that would distinguish them from the legacy service providers. Although there has been virtually no service differentiation between the new CLEC competitors and the ILECs and in the aggregate they still have only managed to account for about 15% of the gross revenues in the space, the ILECs maintain that they should now be allowed to proceed unfettered from any regulatory mandates that were supposed to create a sustainable competitive environment.

Further exacerbating the failure of true facilities–based competition to emerge was the unfortunate collapse of the financial market support for preprofitability telecommunications competitors after only about three years of rapid expansion and building of new infrastructure. This abandonment of the sector by the institutional financial community so early in its infancy resulted in an environment where the survivors among the pure telecommunications competitors now exist as weaker companies, post-bankruptcy restructuring, with the new owners seeking to achieve near-term profitability off the limited installed base of facilities, rather

---

* Report and Order Implements Infrastructure Sharing Provision of the Telecommunications Act of 1996. Dkt No.: CC-96-237. Adopted: February 6, 1997.

than pursuing autonomous access network expansion. Cautious optimism, slow but steady growth, and continued reliance on the ILECs for the majority of their access circuits mark the present state of competitiveness among most of the CLEC industry. It is in this environment that wireless broadband services of all types are poised to reemerge as the most logical and cost-effective means for competitors to the existing wired networks held captive by the ILECs and the cable companies.

A lot of smoke and excitement was recently generated by the Broadband over Powerline (BOPL) technologies, giving false hopes to the regulators and policy makers that indeed there was a third path into the homes and offices to provide worthy competition to the ILECs. To date the BOPL trials have proven to deliver mixed results, with little to be gained in terms of either performance or cost savings compared with alternate technologies, including fiber and wireless access technologies.

Our industry is at a crossroads, where the decisions to compete for the growing basket of services, which consumers and businesses are able and willing to purchase, now encompass purchases not only for voice and data but also for television and enhanced services.

We have failed as an industry to achieve anything approaching a level playing field among the incumbent service providers and the ever-hopeful competitors who thus far have survived off the table scrapes of the major players. True competition will only happen when competitors realize that anything short of end-to-end ownership (or contractual control in outsourced situations) of their service delivery networks will enable them to finally compete on at least a fairly level field of engagement.

RADIO SPECTRUM ALLOCATIONS

Of central interest to the expansion and enhancement of the wireless industry is what has evolved as a complex and painfully slow process for licensing radio spectrum to operators and erstwhile service providers or private network users. The most important issues to address in the radio spectrum allocation and management arena include the following:

1. The need for frequency licenses to be technology neutral.
2. The need for substantial increases in the amount of bandwidth available for licensing to enable broadband wireless access networks in the frequency ranges from 450–6000 MHz, including both licensed and unlicensed spectrum for both commercial and public safety applications.
3. The allocation of spectrum in sufficient contiguous blocks to enable TDD networks to exist free from the radio interference that happens when trying to coordinate services with the traditional bias (and the majority of the mobile spectrum allocations on the existing frequency table) with paired FDD allocations. Regulators have not yet caught on to the imperative for eliminating the distinctions between FDD-paired channels, which are optimal for high speed mobile networks, and TDD spectrum, which is optimal for converged (fixed, portable, and mobile) broadband data

communications. If the broadband wireless networks that are essential to the growth of our market are to be realized, additional support of unpaired TDD spectrum is essential. The distinction will have to unwind over a sufficient period of time to allow for the coexistence of legacy FDD and new network TDD technologies. There is a growing requirement for literally hundreds of megahertz of additional spectrum to be allocated for next-generation services at frequencies below 3 GHz.

4. Refarming of legacy spectrum allocations to better align the spectrum with modern digital radio systems, and especially broadband networks. Historically, radio frequencies were assigned to specific market applications (maritime, aeronautical, private land mobile, auxiliary broadcasting, timber, taxis, etc.) and specific radio technologies such as fixed, mobile, paging, and push-to-talk.

5. Increases of license-exempt spectrum allocations and increases in their allowable power output for rural systems.

6. Expansion of the concept of the "lightly licensed" spectrum as seen in the recent rules adopted for the 3.65- to 3.70-GHz bands. This concept rewards licensees who are the first to construct their facilities in a given area, requiring subsequent licensees to coordinate with those who are in service before the subsequent entrants into the licensed service area, but which do not prevent subsequent entrants.

Taken together, these developments are contributing to the realignment of the competitive telecommunications industry and heavily influencing the next wave of wireless broadband services delivered over converged networks owned (or outsourced) by the cellular, cable television, satellite television, independent telephone companies, and Internet service providers.

RADIO SPECTRUM AUCTIONS: A FAILED POLICY?

Depending on the objectives of the policies imposed on the industry by the regulators with oversight responsibility, it is sometimes hard to determine the successful implementation of a policy and one that is often marred by the law of unintended consequences.

In the authors' opinion, the spectrum auction policies followed by the FCC to place radio spectrum into the hands of groups who will exploit these assets for the best and highest good of the public falls far short of this mark. Present FCC auction policies evolved as just the latest step along a continuum of previously flawed policies regarding spectrum licensing and allocation. As wireless spectrum-based businesses proved to hold substantial value in the market, there emerged a frenzied quest by Congress and the FCC to collect auction revenue. This enthusiasm was initially fueled by the relatively large values bid by the industry in pursuit of the original PCS auctions. In addition, other frequency bands were soon subject to the auction enthusiasm as the FCC sought ways to ensure that the triggering event, mutual exclusivity, was present in virtually every band that came up for license.

Unfortunately, as always happens when we paint with a broad brush, all the nuance is about the actual differences between many different types of wireless services, the types of groups that were able to prevail in the various auctions, and their interest in actually deploying services of value in a timely manner to the public. It is becoming clear in hindsight that many of the prevailing bids for spectrum purchases over the past several years were made by existing service providers whose primary interest was in warehousing spectrum so others could not use it to compete against them or, alternatively, by spectrum real estate speculators whose interest is not in deploying services but in optimizing financial returns from subsequent private sale of the spectrum assets. The proof of this observation is the recent outcome of the highly valued and unique spectrum in the 700-MHz bands that was almost entirely fetched by the leading incumbent operators in the recent FCC auction number 73.

Lost along the way to our current situation were the dramatic engines of economic growth fueled by creative entrepreneurs who were able to acquire spectrum through first-come, first-served applications or via methods of allocation that have been abandoned along the way by the FCC's seeming inability to create a level playing field between wealthy incumbents and financially strapped new entrants. The concept of the best and highest use of radio spectrum resources that is in the public interest has been lost through a largely thoughtless and suboptimal auction mentality.

Even if auctions are the best we can do to achieve the public interest in spectrum allocations and licensing, we need to seek ways to increase our creativity and thoughtful analysis and reflection on how the spectrum allocation process can be improved. Some alternative approaches to the currency being bid should be thoroughly evaluated; for example, rather than a onetime upfront payment, which is highly punitive to service providers, they must finance the purchase cost *and* the expense of network installation on the front end of any services being delivered. Thus, bids presently reflect a discount to the ultimate value of the future earning potential of the business. An alternative that could be entertained is the concept of bidding a revenue sharing percentage. Another "bidding credit" might be how soon the service will be deployed post receipt of license. Long periods of inactivity are presently common for the perfection of licenses and a fairly liberal waiver and extension mentality by the commission to allow license holders extensions over many years before they actually implement operations. If the FCC was true to its charter, then milestones for rollout could also be another form of currency (or punitive reclamation tool) to enhance the spectrum auction process. These alternative approaches would augment the existing preferences and set-asides for minority and small businesses and other policies that could be created to equalize the disparity between the creative cash-poor innovators and the established major players in the industry.

Regardless of how the present inefficient situation is resolved, it needs to be highlighted that we are slipping steadily farther behind our international brethren with the implementation of advanced wireless systems.

FINANCIAL REALITIES

There are a wide range of financial issues and by-products of our present telecommunications evolution. The U.S. economy has enjoyed a relatively stable contribution of large top-line benefits from the telecommunications industry over the past century. Local tax collections from both service revenues and property taxes are likely to become eroded as the top-line revenue and capital expense base of the telecommunications industry contracts.

Reductions in the large number of employees that have historically been required to deliver telecommunications services is also reducing as automated systems and outsourced off-shore call centers contribute to a negative trickledown effect through the economy. Thus, as the scale of the industry contracts as an employer and creator of income tax revenues and the corresponding benefits of the employees' purchasing power are reduced, there will inevitably be some painful fallout that hits far and wide in the industry.

Following is a list of some of the leading issues that must be carefully considered by the policy makers, managers, executives, investors, and entrepreneurs who will be tasked with managing a major change in the structure, scale, and revenue of the telecommunications industry broadly defined.

- Potential for volatility among traditional industry financial leaders
- Protection of pensions
- Wide range of economic damage potential
- Tax base erosion
- Capital expense burden
- Rapid obsolescence/technology upgrades

As we enter the next phase of broadband proliferation and the continued growth of the Internet into a media-rich, broadband-consuming environment, the risk of financial volatility erupting at various stages throughout our industry will continue apace with each development of a new disruptive force that emerges out of the search for improved efficiencies and profitability at the expense of some highly valued, but soon to be obsolete, legacy institutions and market leaders. Voice services are at the top of our list of highly valued, well-established services, currently provided by top-heavy organizations that are ill equipped to grapple with a reduction in their gross revenue streams of between 70% and 85% over the not-so-many next few years. How the legacy telephone companies and their cellular operations cope with this inevitable reality is the 900-pound gorilla sitting in every telecom analysts' waiting room, with Skype and its brethren breathing down their proverbial necks.

As Moore's law continues to capture traction in the wireless communications industry and our devices and network infrastructure equipment continuously grow in sophistication, the advances in computing power and efficiency will continue to reduce costs and increase functionality. Unfortunately, our technical progress is too often wasted on uninspired knockoffs of the most popular phones of the recent period, or alternatively, on expensive, powerful, yet arcane features and user interfaces. In essence, as an industry, we are often squandering the incredible blessings of Moore's law on the trivial and repetitive, failing to leverage increased device power to expand the utility and ease-of-use functionality that leads to true sustainability in the mass market. This insidious march of progress plays havoc with just about everyone participating in the industry, carriers are subjected to an endless stream of upgrades and capital expense requirements, hardware and software vendors fight against early obsolescence, squeezing their time to recover non-recurring development expenses, and customers become polarized into either the demanding and understanding enthusiast camp or the frustrated and often angry majority of the uncertain and uninformed. Criticism is cheap, and self-flagellation is best left to religious fanatics, so where do we recommend turning to for improving our products and the economic performance and survivability of our service providers and especially all the interrelated participants in the supply chain that feed the ecosystem of the wireless broadband world?

We hope to provide some insights for those responsible for making business and technical decisions regarding the pace and direction of change within the confines of the wireless broadband sector. Our investment and technical decisions are so torturously intertwined that any wrong step will inevitably be a career-ending decision for those who take their organizations in directions that subsequent hindsight illuminates as faulty leadership. How we make decisions in large, midsized, and small organizations is a topic that literally fills large sections of libraries. However, to avoid the high risk decisions our leaders are paid to make in favor of careful incrementalism, placing excess focus on the short-term financial performance of our organizations will come at the expense of our preparedness for growth, evolution, and even survival in the rapidly arriving broadband future.

Will the AT&T and Verizon executives have the heart and stomach to realize that the flywheel of voice that has driven their financial engine so powerfully for the past 100 years is about to slow down precipitously? Can they step in front of the analysts who track them with a coherent story that rings of hope and promise for a strong future of growth in bandwidth and enhanced services and a healthy environment of open access to enable and empower new applications developers to have a ready path to market? Verizon's leadership has demonstrated its commitment to investing in both its fiber-to-the-home FiOS project, expected to reach approximately 7 million video subscribers, 11 million Internet subscribers by 2010, and earnings before interest, taxes, depreciation, and amortization (EBITDA) positive by 2009 [1]. However, the pace of this buildout and its financing is based in large measure on the voice revenue engine, which we predict is at some substantial risk over the coming few years, as broadband proliferates into homes and offices and dedicated voice circuits are migrated to shared packet data networks.

AT&T on the other hand has been far more conservative in its pursuit of network upgrades on both the wired and wireless sides of the equation. Its WCDMA 3G deployments cover far less territory than Verizon has achieved with its EV-DO wireless data overlay network. Further, AT&T's video strategy is less aggressive, with more modest objectives published thus far than the announcements of Verizon. As a result of the challenges inherent in digesting all the recent merger activities that have combined to restore the AT&T brand name to the status of the "biggest" telecommunications company, it appears that AT&T has failed to provide any visibility into how it expects to become the "best" player in the industry. The authors suspect that the company will mount a tortuous "rearguard" action against progress, much in keeping with the actions of its former piece part organizations over the past decade before reconsolidation. Alternatively, will it move more aggressively with its present fiber-to-the-premise projects and expand them to become a true national initiative, which will position the legacy telephone companies for a leadership role in the broadband future? The major decision facing its management is their willingness to accept short-term negative impacts on earnings, while positioning to lead in the near future.

In closing, the authors predict that if they fail to evolve their access and transport networks into next-generation broadband packet converged networks, with fiber to the premise of the majority of their customers, quickly, then they will have lost the leverage of their legacy voice revenue streams, which will result in a transformational devolution of their market position. Failure to self-cannibalize revenue and services over upgraded broadband fiber and wireless networks in pursuit of maintaining the status quo is a recipe for disaster. The victims of this disaster will be widespread among all the stakeholders, including management, employees, shareholders, the customers, the public, and the nation at large. The authors are neither contrarians nor anti-ILEC; we are, however, uniquely exposed as industry observers and participants to a wide range of external factors that may simply be incomprehensible to the decision makers in virtually every corner of the impacted ecosystem surrounding our largest legacy service providers. If we insist that the emperor has "new clothes," then prepare to be surprised, but if we realize the naked, transformational power of the broadband Internet and the proliferation of wireless broadband services into new markets, then maybe we can meet the future with fewer disruptive bumps along the way.

THE STANDARDS WARS: PROPRIETARY VERSUS OPEN STANDARDS

Before the trade associations or forums that govern interoperability and promote specific standards-based products into commercial service, there is a long and often contentious process of standard development. In the electronics industry, the IEEE provides the oversight and manages the governance and rules of engagement for the contending solutions seeking to garner a standard designation for its proprietary product or solution. The standards-setting process begins with the receipt of a petition for the creation of a working group to manage the process of evaluating all submissions for inclusion in the standard defining process once the IEEE charters a working group to manage the process.

THE MANY FACES OF THE STANDARDS PROCESS

- When open standards are essential
- When proprietary standards prevail
- Common air interface standards success stories
- Network protocols

When Open Standards Are Essential

Standards are essential to those products and services that require multiple vendors to simultaneously participate in the delivery of the intended service. Wireless cellular systems are especially compelling models for standards, as each layer of the infrastructure will likely be provided by different companies with widely diverse skills and expertise. Companies that concentrate on core network equipment most often manufacture base stations and switches. Handsets, on the other hand, are manufactured by a wide range of suppliers, and it is the service provider's hope that as many alternative vendors and choices of user terminal hardware are available. So it is the creation of standards for the "common air interface" that is essential, enabling any manufacturers that want to build conforming products to know from the technical specifications devised for the standard exactly how to manufacture the interoperable capabilities of their products. Alternatively, point-to-point radios are seldom built to standards on the air-link because they will inevitably be "talking" with only other devices supplied by the same manufacturer. However, even in these cases, the interconnection and message traffic–handling capabilities and network connectors will inevitably be conformed to international standards for such interfaces.

Standards have proven to be a major contributor to the mass-market growth of many of our treasured services, including television, radio, cellular phones, and virtually the entire telephone network linking the globe. Throughout the history of the electronics industry, the development and implementation of industry standards have been undergoing a steady evolution and refinement process. Originally set by fiat either by pioneering engineers or by government mandate, the standards process has democratized in recent years to embrace an open process of engagement and refinement from all of the interested parties who wish to participate.

On the negative side of the current approach is the inevitable dilution for some of the most innovative contributions, as they become homogenized into a process marked by compromise and the "big-tent" approach that attempts to incorporate as much of the content proposed from the entire swath of participants. The larger the market opportunity, the greater the attention paid to the process by more participants and the resulting offset to proprietary approaches and submissions.

There are also numerous opportunities for vested interests to sabotage or hijack the process under the current standards-setting protocols.

When Proprietary Standards Prevail

Also ever present are the market forces that reward owners of proprietary solutions and products that gain enough market traction early enough in their development life, that they are able to largely avoid the dilution inherent from the incorporation of their intellectual property rights (IPR) into a standard. The most famous of the beneficiaries of this alternative "nonstandard" approach is Microsoft. A virtual monopoly position in personal computing has been captured by this single company's proprietary operating system and application software. However, the success of the Microsoft suite of products has over the years depended on Microsoft's embracing a wide range of standardized interfaces and communications protocols into its application suite of software products. Thus, even the most successful nonstandard product organizations are faced with the requirement to embrace a large number of standards that intersect with their products.

The complex area of cross-licensing of intellectual property between and among the companies that have developed specific IPRs, who then seek to have their IPR embedded into standards, is a controversial and challenging environment that overhangs the standards-setting process. The translation of a standards process into a commercial licensing business has been most famously accomplished by Qualcomm, which developed most of the core technology used in CDMA cellular equipment and subsequently proposed its incorporation into the standards established for CDMA cellular systems (IS-95).

Companies that seek to manufacture products conforming to various standards that have proprietary IPR embedded in the standards face four primary approaches to licensing the IPR in question.

1. First, they may have to seek direct license and royalty payment terms with the owner of each governing patent—a cumbersome, expensive, and time-consuming process.
2. Second, they may be in a position to cross-license with the subject patent holders in exchange for cross-licensing rights to IPR that they may also own and that other patent holders may find helpful to their own business objectives. In these cases, a mutual release is often negotiated and no money or royalty payments are required. This was the original settlement between Qualcomm and Ericsson and Nokia for cellular patents.
3. Third, industry groups with overlapping patent interest in various standards may often create a "pooling of interest" approach, under which all the holders of applicable intellectual property to a given standard combine their interests into a single license granting and royalty payment collecting entity. This approach has been most visibly applied in the GSM standard group, wherein there are a very large number of patent rights holders who all seek participation in the licensing revenue.

4. The fourth regime is one that will soon face the test of fire in the WiMAX marketplace, and that is the mandate by the standards-setting group that all participants in the process agree to cross-license their IPR to other organizations working to build products under the standard, under "market standard, commercially reasonable terms." Unfortunately, the precise definition of these terms have been left to the imagination of future negotiators and the decision of lawyers, which makes the whole concept pushed by Intel of a patent licensing–free WiMAX standard to be a questionable objective at best. Time will tell how the massive amount of IPR that is incorporated into the 802.16(e) standard will be rationalized by the standards group, the trade association (WiMAX Forum) and among the large number of rights holders that are also involved with the WCDMA LTE product development. Largely ignored to date are the core system level OFDMA patents owned by Adaptix, Inc., which were earlier developed by Dr. Hui Liu and his team.

Proprietary Standard Success Stories

Microsoft
Phillips CD
Apple iPod

Common Air Interface Standards Success Stories

GSM and CDMA cellular
Ethernet IEEE 802
WiFi IEEE 802.11(b)(a)(g)(n)
WiMAX IEEE 802.16(d)(e)

Network Protocol Standards

The Open Standards Interconnection seven-layer protocol stack (Fig. 12.1) developed by the International Standards Organization is yielding to the ubiquity and greater simplicity of the Internet using the TCP/IP protocol suite (Fig. 12.2) and its derivatives developed by the IETF*.

The IP suite is the set of communications protocols that is the core of the Internet and Ethernet communications based on the TCP and the IP, resulting in the full TCP/IP protocol suite.

TCP/IP has eclipsed literally every competing core network technology for dominance in the NGN evolution. It is central to the evolution of wireless broadband core networks, as they evolve from traditional circuit-switched

---

* Internet Society International Secretariat, 1775 Wiehle Ave., Suite 102, Reston, VA 20190.

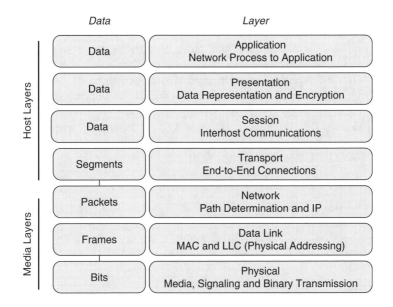

Figure 12.1. Open Standards Interconnection seven-layer stack.

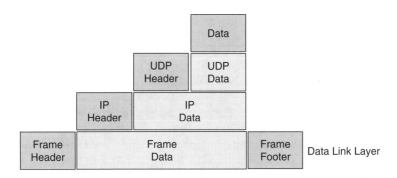

Figure 12.2. The "Internet protocol suite" or "TCP/IP protocol suite."

architectures to the full packet-based network topologies of the immediate and foreseeable future.

It is a living standard created under the umbrella organization called the Internet Society (ISOC). The Internet standards are under constant review and enhancement initiatives under the ISOC auspices via the IETF, which is an international open forum comprising network engineers, system designers, service providers, infrastructure vendors, academics, and researchers cooperating with the constant development and evolution of the Internet architecture and its operations.

The IETF is organized into a wide range of working groups concentrating on specific topics, including, for example, routing, transport, security, protocol modifications, and architectural enhancements to the Internet. The IETF holds meetings about three times per year, with progress of the working groups managed via e-mail between and among the participants, under the coordination of area directors (ADs). Area directors are members of the Internet Engineering Steering Group (IESG).

The Standards Process: The Games People Play

Standards working groups are incredible organizations, and the participants attend for a wide range of reasons personal to their own business objectives. Depending on their goals and objectives, their participation within the working group may evidence a wide range of behaviors.

- Offense
- Defense
- Obfuscation
- Delay
- Protectionism
- Greed
- Altruism
- Political aims
- Industrial policy objectives
- Government public policy positions (China, Japan, United States)

The standards-setting process is very much like the metaphor about making sausages. Those nice, tidy plastic-wrapped products set out for us in neat order came from a process that was truly not very "pretty." The IEEE process is not easy, it is expensive to participate in, with little guarantee that the contributor's creativity will result in much more than the satisfaction of being a good citizen of our engineering community. However, the benefits to the public far outweigh the effort required to finalize a standard.

REFERENCE

1. Verizon Q4 Earnings Report. Press Release; Jan. 28, 2008.

13

CONCLUSION

The ITU asserts that access to affordable telecommunications is a human right. Politicians all over the world are trying to bridge an emerging digital divide between the well-served metropolitan area residents and the underserved residents of rural communities. In the successful information society, the key communications network infrastructure must be as ubiquitous as possible to include all persons into the empowering data and communications resources. If nations fail to bridge the broadband digital divide, there will be increasing asymmetry between the participants in the information society versus those with reduced access to information resources by populations in lesser-developed regions. A significant by-product of this situation will be the disruptive forces of financial growth in the affected areas and the inevitable population migrations to better connected urban areas. Equally challenging are the structural and political problems that will be caused by increasing gaps in education and rising unemployment, leading to the need for increased subsidy payments to rural areas.

The key driver of future economic growth in the information society will be the participation rate of the citizens in the information infrastructure. Numerous nations, including the United States, Canada, Australia, and India, have recently set national targets for expanding broadband availability among their populations. In the United States alone, the historic universal service funding power (> $2 billion of loan guarantees in 2004) of the Department of Agriculture, under the Rural Utility Service (RUS) program, is now open on a first-come, first-serve

Wireless Broadband. By Vern Fotheringham and Chetan Sharma
Copyright © 2008 the Institute of Electrical and Electronics Engineering, Inc.

basis for new broadband service providers in second- and third-tier markets of up to 20,000 in population.

Similar national objectives exist across the world. The World Bank Group is working on a supranational level to ensure that the information society benefits also reach the developing countries, which in the postindustrial era will also fuel economic growth in the developed nations. The inclusion of all global economies into the future interconnected broadband world is both essential and ultimately inevitable. This goal is well within our reach if we continue to expand the evolution of the global cellular networks from narrowband voice-centric platforms to broadband Internet wireless networks. How soon and under what commercial terms and conditions this future will be reached constitute the stakes of the current conflict we have been examining throughout this book. It is inevitable that the expansion of wireless broadband access networks will continue through periods of both boom and bust, the trend is insidious and unstoppable. However, it will only be from the hindsight of the future that our generation of pioneers will be judged for our wisdom, competence, and creativity in solving the numerous financial, regulatory, and technical challenges facing the creation of the universal information society on a global scale.

ECONOMIC GROWTH

The stratification of access and transport networks under NGN infrastructure is going to provide enormous opportunity for creating new markets and applications, and the opening of previously closed networks will, we believe, expand the value chain and lead to increased profitability to all participants within the ecosystem of the broadband wireless networks of the future.

With high capacity fiber extending out to individual homes and businesses, the need to add the utility value of mobility and portability of those broadband-fixed services will drive the adoption and expansion of the broadband wireless Internet into virtually every corner of the world. The revolution under way in the advertising industry is noteworthy, as interactive tools can accurately measure the effectiveness of a specific advertisement to reach its intended audience and contribute directly to its purchasing behavior. This ability to increase efficiency and effectiveness of information delivery adds tremendous value to all participants in the transaction, resulting in higher profits for vendors and better service and lower costs to consumers.

The Internet was originally lauded for its ability to count "eyeballs"; the search revolution led by Google has shown how to monetize "eyeballs," and the recent incorporation of feedback from consumers to specific advertising messages has completely broken the mold on the traditional economics of advertising and distribution. The contributions to the global economy from this dramatic increase in commercial commerce efficiency will be dramatic and expansive. We are in the midst of seeing how the disintermediation capabilities of the Internet, when augmented and enhanced by the addition of information both to, from, and about the consumer and the vendor, add value to specific transactions and the economy in general. Who wins and who loses among the participants in the traditional retail and wholesale distribution chains is beginning to emerge, as high overhead retail establishments fight to maintain market share by morphing into entertainment and sales venues, relying on customers' physical experiences to offset the pull of the growing virtual retail space offered by the Internet. The conflict is far from over, but the battle lines are becoming clear. There is a contest underway between the efficiency of the virtual web-based marketing environments versus the experiential benefits of the physical retail environment; the authors cannot help but believe that it will be the combination of both that prevails.

While still on this subject, we should also examine the broader concept of learning and education. With global access to the best minds and teachers becoming available via the Internet, the expense and inefficiencies of traditional educational institutions are soon in for a major challenge from the virtual realms of the broadband Internet versus the physical benefits available to students' physical attendance in a campus setting. Again, we predict that the outcome will indeed be a combination of both environments. Gaining access to learning via access to information will inevitably tilt toward the Internet, but the need to learn essential socializing and human interactive skills that are only available through the direct physical contact with teachers, mentors, and peers will also require immersion in the physical domain of education.

PUBLIC POLICY

We will all enjoy the benefits of the broadband future, yet there is much work to be done, with many regulatory and technical battles still to be fought over exactly how to bring these services into being. We will get the future we deserve, on the basis of our courage and wisdom in how we balance massive, highly disruptive change, without destroying much of the value created in the narrowband and wideband eras, which have enormously benefited our society at large, and each of us personally, as we move forward into the broadband era.

We challenge our politicians, regulators, financiers, carriers, vendors, and the consuming public to pay very close attention to how we balance the public policy issues and the sometimes-painful business impacts that are expected. Too heavy a hand in any one dimension of the debates will surely have negative impacts on overall progress. How the pains and the gains are shared and distributed across all boundaries of the industry and its publics will determine our ultimate success, or alternatively, just subject us to a muddled mess that ultimately penalizes everyone, as we lose our competitive advantages as a global leader and innovator. Will the United States regain its position as a world leader in advanced telecommunications, or will it slowly but surely slip into the role of a tired old economy, past its prime and slipping into its declining years? The potential to do damage is real, and the potential for the United States to lose its position as a world leader is also real and imminent. We presently live in a time of highly polarized political agendas.

Surely, broadband policies and the extension of U.S. international competitiveness is a topic that affects all political camps equally. We urge the agenda setters in Washington, DC and at the state levels to actively pursue a high profile public policy debate regarding these matters. We also urge that the powerful and well-oiled influence machine of the major incumbents be held in check by the new activism of the Internet and computing industries, including such firms as Google, Microsoft, Yahoo!, IBM, Dell, HP, and their peers. At the same time and under the same framework, we should also include the debates over digital rights management, patent, and copyright protection into the mix, as these are central to ensuring that our inventors and artists will be incentivized to creatively address and be rewarded for developing the complex ecosystems and content that are required to enable the fully converged future.

Good luck to us all, and we look forward to seeing you in the future, literally anywhere and at anytime, using our emerging wireless broadband infrastructure and services.

A

WIRELESS BROADBAND GLOSSARY

1xEV-DO (One Carrier Evolved, Data Optimized): The description of the 3G data only solution from Qualcomm that delivers 2.4 Mbps per sector. Typical user throughput is between 300 and 600 Kbps. Revision A to 1xEV-DO provides increased downlink data rates to 3.1 Mbps and increases the return path to 1.2 Mbps.

1xEV-DV (One Carrier Evolved, Data Voice): The Qualcomm 3G solution that incorporates voice and data on the same carrier with rates up to 4.8 Mbps.

1xRTT (One Carrier Radio Transmission Technology): The description of the Qualcomm CDMA cellular platform using 1.25 MHz channels, sometimes called narrowband CDMA.

2G (second generation): The description of the initial digital evolution of cellular technology past the original analog technologies.

3G (third generation): The description of the enhanced evolution of digital cellular to support higher data rates and multimedia services.

3GPP (Third-Generation Partnership Project): Defines the collaborative efforts of regulators and industry to establish the 3G mobile phone standard. The 3GPP specification is under the umbrella of the International Mobile Telecommunications—2000 project of the International Telecommunication Union. 3GPP specifications are based on evolved Global System for Mobile Communications specifications. The participants include the European Telecommunications Standards Institute, Association of Radio Industries

Wireless Broadband. By Vern Fotheringham and Chetan Sharma
Copyright © 2008 the Institute of Electrical and Electronics Engineering, Inc.

and Businesses/Telecommunication Technology Committee (Japan), China Communications Standards Association, Alliance for Telecommunications Industry Solutions (North America) and Telecommunications Technology Association (South Korea). 3GPP should not be confused with Third-Generation Partnership Project 2, which defines the standards based on IS-95 (CDMA), commonly known as CDMA2000.

3GPP2 (Third-Generation Partnership Project 2): The follow-on working group that is focused on the Qualcomm solution called CDMA2000.

4G (fourth generation of cellular radio systems): Describes what will be the ultimate implementation of the OFDMA-based technical standards that are deriving from the GSM LTE and Mobile WiMAX platforms.

8-PSK: octagonal phase shift keying.

AAA: authentication, authorization, and accounting.

AAS: Adaptive Antenna Systems.

AGW: access gateway.

AMR: adaptive multirate.

AMPS: Advanced Mobile Phone System.

ANSI: American National Standards Institute.

API: application programming interface.

ARQ: automatic repeat request.

ARPU: average revenue per user.

ASIC: application specific integrated circuits.

ASP: Application Service Provider.

ATC: ancillary terrestrial component.

ATCA: advanced TCA.

ATL: application transport layer.

ATM: asynchronous transfer mode.

AWS: advanced wireless services.

attenuation: The loss of signal strength that occurs as a wireless signal travels through the air and past (or through) obstructions located within the Fresnel zone. See also **path loss, Fresnel zone**.

AVG: advanced video coding.

AWGN: Additive White Gaussian Noise Channel.

azimuth: Defines the radial coverage of an antenna system. In radio systems azimuth defines the horizontal plan of coverage from an antenna lobe.

bandwidth: A term that has two applications. First, it describes the amount of radio spectrum employed in a given communications channel. Most commonly described in terms of kilohertz or megahertz of bandwidth. Second, it is often used to describe the raw data rate of a digital communications channel, as in the system has 10 megabits of "bandwidth."

BCCH: broadcast control channel.

beamwidth: A description of the physical shape and boundaries of a radio signal created by an antenna. Directional antennas and electronic beam forming antennas have the ability to focus radio frequency emissions into a predetermined shape. These beam patterns have both horizontal and vertical boundaries that define the beamwidth of the antenna. Beamwidth is most often referred to in terms of degrees off the emitting antenna. See also **azimuth and elevation**.

BOL: Broadband over Powerline.

bps: bits per second.

BRS (Broadband Radio Service): Formerly known as the Multipoint Distribution Service/Multichannel Multipoint Distribution Service, is a commercial service. In the past, it was generally used for the transmission of data and video programming to subscribers using high-powered systems, also known as wireless cable. The BRS is now under use as the Mobile WiMAX spectrum bands used by Sprint.

BSC: base station controller.

BT: British Telecom.

BTS: base transceiving station.

C/I: carrier to interference ratio.

CAD: computer-aided design.

CAM: computer-aided manufacturing.

CAPEX: capital expenditure.

CATT: Chinese Academy of Telecommunications Technology.

CBU: Cello basic unit.

CCD: charge-coupled device.

CDF: cumulative distribution function.

CDMA: code division multiple access.

CDPD: Cellular Digital Packet Data.

CMMB: China Mobile Multimedia Broadcasting.

CMOS: complementary metal oxide semiconductor.

CODECS: compression coder-decoders.

CP: cyclic prefix.

CP-OFDM: cyclic-prefix—orthogonal frequency division multiplexing.

CPC: continuous packet connectivity.

CRM: customer relationship management.

DAB: digital audio broadcasting.

dB/decibel: A logarithmic representation of a ratio that expresses gain or loss.

dBd: decibels with respect to a dipole (basic antenna element).

dBi: decibels with respect to an isotropic radiator (antennas).

dBm: decibels with respect to one milliwatt.

dBW: decibels with respect to one watt.

DEMS: digital electronic messaging system.

dipole: A basic antenna element to which most other antennas are compared for gain.

directional antenna: An antenna configured to focus radio spectrum energy in a single direction. Parabolic reflectors and Yagi antennas are examples of directional antenna form factors. See also **omni-directional antenna**.

DSL: digital subscriber line.

DSLAM: digital subscriber line access multiplexer.

DSP: digital signal processor.

DSSS: direct sequence spread spectrum. A spread spectrum signal with a wide bandwidth and low amplitude. A DSSS signal on a nonspread spectrum receiver appears to be noise.

DTM: dual transfer mode.

D-TxAA: double transmit adaptive array.

DVB: digital video broadcasting.

DVB-H: digital video broadcasting–handheld.

DVB-SH: digital video broadcasting–satellite handheld.

DVP: digital video broadcasting.

E-DCH: enhanced dedicated channel.

EBCMCS: Enhanced Broadcast Multicast Services.

EBS (Educational Broadband Service): Formerly known as the Instructional Television Fixed Service, EBS is an educational service that has generally been used for the transmission of instructional material to accredited educational institutions and noneducational institutions such as hospitals, nursing homes, training centers, and rehabilitation centers using high powered systems. Our recent revamping of the EBS spectrum will now make it possible for EBS users to continue their instructional services utilizing low power broadband systems while also providing students with high speed Internet access.

EDGE: Enhanced Data Rates for GSM Evolution.

EGPRS: Enhanced General Packet Radio Service.

elevation: The vertical angle and coverage defied in degrees for an antenna system.

eNodeB: Evolved Node B.

EMS: electronic manufacturing services.

EPS: Evolved Packet System.

ERP: enterprise resource planning.

ERP (effective radiated power): The ERP is equal to the transmitter output power, minus the power loss in the coaxial cable, plus the margin added from the gain of the antenna.

ESN: electronic serial numbers.

Ethernet: IEEE standard 802.3—the dominant standard defining the MAC and physical layer frame–based local area network protocol.

ETRI (Electronic and Telecommunications Research Institute): The Korean government- and industry-sponsored technology development organization.

ETSI (European Telecommunications Institute): The standards body for the EU.

EURESCOM: European Institute for Research and Strategic Studies in Telecommunications.

E-UTRAN: Enhanced UMTS Terrestrial Radio Access Network.

EV-DO: Evolved, Data Optimized.

EV-DV: Evolved, Data Voice.

fade margin: The amount of "extra" signal strength above the minimum needed to establish a given wireless communications path. Fade margin is engineered into network system design to ensure reliable and robust connections are maintained as users are exposed to the ever changing fades and nulls common to mobile communications systems.

f/b ratio (front-to-back ratio): The power ratio (in dB) of the power going out in the favored (front) direction of an antenna compared to the power that leaks out of the back of the antenna. A typical directional antenna will have a f/b ratio of 20 dB or more.

FCC: Federal Communications Commission.

FDD (frequency division duplex): FDD-based wireless systems use separate radio channels for the upstream and downstream transmission. It has been the primary architecture for mobile wireless systems.

FDM (frequency division multiplexing): Combining multiple radio frequency channels into a combined virtual bearer channel.

FDMA (frequency division multiple access): FDMA systems divide the radio spectrum into small channels, such that in a pure FDMA analog system, each user is allocated a dedicated portion of bandwidth for their session. When FDMA is hybridized with the digital domain, OFDMA results, wherein both the spectrum access and the digital and time domains add additional flexibility and control.

FFT (fast Fourier transform): An algorithm used in digital signal processing to create radio frequencies and related applications.

FHSS (frequency-hopping spread spectrum): A spread spectrum signal that uses fast hopping between relatively narrow channel bandwidths to randomly avoid sources of interference or degrading portions of the bandwidth.

Flash OFDM: Fast Low-Latency Access with Seamless Handoff OFDM.

FLO (Forward Link Only): As in Qualcomm's MediaFLO platform.

FMC (fixed mobile convergence): The utilization of wireless networks as replacements for traditional wireline services.

FP7 (Seventh Framework Program): The organization chartered to promote and enhance technology development and excellence within the EU. It is the primary funding source for R&D between 2007 and 2013.

frequency: Defines the number of cycles at which a given portion of the electro-magnetic spectrum oscillates. The phenomenon is named in honor of Hedrick Hertz, who first described the behavior. Each cycle or wave is defined as one Hertz. The wireless systems in common use today are most often defined in terms of MHz (megahertz—millions of cycles per seconds) or GHz (gigahertz—billions of cycles per second) or Hz (Hertz). One thousand Hertz per second is one kilohertz (kHz). One million Hertz each second is one megahertz (MHz). One billion Hertz each second is one Gigahertz (GHz). See also **Hz**.

Fresnel zone: Radio waves disperse after they leave the antenna element. The Fresnel (pronounced "fre-nel") zone is the area along the radio path over which the signal spreads.

FTP (File Transfer Protocol): The IETF standard for file transfers over the Internet.

FTTH: fiber to the home.

FWA: fixed wireless access.

G-Rake (generalized rake receiver): Functions like an equalizer, suppressing self-interference while significantly improving data throughput and system capacity for HSDPA on WCDMA platforms.

Gbps: gigabits per second.

GERAN: GSM EDGE Radio Access Network.

GGSN: gateway GPRS support node.

GHz (gigahertz): one billion (one thousand million) Hertz. See also **frequency**.

GMSK: Gaussian minimum shift keying.

GPRS: General Packet Radio Service.

GPS: global positioning system.

GSM: Global System for Mobile communications.

GSMA: GSM Association.

HARQ (hybrid automatic repeat request): An enhanced version of ARQ, the forward error detection and correction technology. It is used in wireless environments where the potential for disrupted signals is common. It is incorporated in the 3GPP standard HSDPA and HSUPA, and also for mobile networks such as UMTS, and Mobile WiMAX.

Hz (Hertz): One complete cycle of a wireless signal. See also **frequency**, **kHz**, **MHz**, **GHz**.

HLR: Home Location Register.

HSDPA (High Speed Downlink Packet Access): The WCDMA high speed data enhancement for downlink data transmission.

HS-PDSCH: High Speed–Physical Downlink Shared Channels.

***HSPA* (High Speed Packet Access) (HSDPA with HSUPA):** WCDMA enhancements for increased data transmission speeds and capacity.

HSPA +: HSPA Evolution.

HSUPA (High Speed Uplink Packet Access): The WCDMA high speed data enhancement for uplink data transmission.

ICT: information and communication technologies.

IESG: Internet Engineering Steering Group.

IXC: interexchange carrier.

IEEE (Institute of Electrical and Electronic Engineers): A professional organization that helps set transmission system standards.

IETF (Internet Engineering Taskforce): The standards organization for the Internet.

IFFT (inverse fast Fourier transform): The reciprocal of FFT calculations.

IFPI: International Federation of the Phonographic Industry.

IM: instant messaging.

IMS: IP Multimedia Subsystem—the packet.

IMT (International Mobile Telecommunications): IMT-2000 describes the six ITU certified standards for cellular communications. They include (1) WCDMA, (2) CDMA2000, (3) TD-CDMA and TD-SCDMA, (4) IMT-SC single carrier (EDGE), (5) DECT, and (6) Mobile WiMAX.

IO: input-output.

IPR: intellectual property rights.

IP: Internet Protocol.

IPTV: Internet Protocol television.

IR (incremental redundancy): An enhancement to the GSM-based Enhanced General Packet Radio System that reduces system overhead by not sending error correction data over the airlink unless an error is detected by the receiver thereby improving efficiency.

ISI: intersymbol interference.

isotropic: A theoretical antenna that would radiate signal equally in all directions. It is used as a reference point against which the gain of a physical antenna is compared (in dBi).

ISP: Internet service provider.

ITFS: Instructional Television Fixed Service.

ITU (International Telecommunications Union): The global organization that oversees all telecommunications standards and coordination between the nations.

JCP (Java Community Process): A cooperative organization to govern the open, participative process for the development and evolution of the Java™ technology.

JPEG: Joint Picture Experts Group.

Kbps: kilobits per second.

kHz (kilohertz): one thousand Hertz. See also **frequency**.

km: kilometer.

LBS: location-based services.

line of sight: A clear wireless "line-of-sight" path describes an environment between both ends of a point-to-point link that has no obstructions between the antennas.

LMDS (Local Multipoint Distribution Service): LMDS is a broadband wireless point-to-multipoint communication system operating above 20 GHz (depending on country of licensing), which can be used to provide digital two-way voice, data, Internet, and video services.

LTE (Long-Term Evolution): The 3GPP cellular road map for the "long-term evolution" of the GSM platform into a fourth-generation system. It is anticipated to include an implementation of OFDMA technology within the framework of the legacy cellular network topologies.

LTU: local timing unit.

MAC: Media Access Control.

MBMS: Multimedia Broadcast/Multicast Service.

Mbps: megabits per second.

MCPA: multicarrier power amplifier.

Mcps (megachips per second): Describes the speed at which "chips" (the encoding elements used in DSSS/CDMA communication systems) are created. The speed generated is called the "chipping rate." WCDMA systems employ a chipping rate of 3.84 Mcps.

MCS: modulation and coding scheme.

MediaFLO: Media Forward Link Only.

MFJ: Modified Final Judgment.

µs (microseconds): one millionth of a second.

MHz (megahertz): one million Hertz. See also **frequency**.

MIMO (multiple-input and multiple-output): A smart antenna solution that uses multiple antennas at both the transmitter and receiver ends of a radio path to improve performance. It increases the amount of bits per second per Hertz of bandwidth, without requiring additional power. This benefit makes MIMO and its derivatives very helpful addition to broadband wireless systems. It has been adopted as integral to the Mobile WIMAX standard and the 3GPP LTE road map for GSM systems. See **SIMO, SISO, MISO**.

MISO: multiple input single output. See also **MIMO**.

MMDS (Multichannel Multipoint Distribution Service): This was the description of the original wireless cable spectrum between 2.5 and 2.7 GHz. It also incorporated the ITFS (Instructional Television Fixed Service) service. The service has been reassigned to the BRS and ERS, which is the spectrum primarily used by Sprint and Clearwire for their Mobile WiMAX systems. See also **BRS, ERS**.

MITF: Japan Mobile IT Forum.

MME: mobile management entity.

MMS: multimedia messaging.

MMSE: minimum mean square error.

MPEG: Motion Picture Experts Group.

MRxD: mobile receive diversity.

MSC: mobile switching center.

msec (millisecond): one thousandth of a second.

multipath: The nearly simultaneous reception of a direct signal and one or more reflected echoes of the signal. Historically, multipath was a significant contributor to degradation of the received signal. Recently, the introduction of MIMO technology into wireless systems is actually leveraging multipath reflections to increase the performance of the wireless channel. See also **MIMO, SIMO, smart antennas**.

MVDDS (Multipoint Video Digital Distribution Service): The terrestrial licensed spectrum in the United States that is coprimary with DBS spectrum.

mW (milliwatt): one thousandth of a watt. See also **Watt**.

NGMC: Next-Generation Mobile Committee.

NGN: next-generation network.

OAM: operation, administration, and maintenance.

OBAN: Open Broadband Access Network.

OEM: original equipment manufacturer.

ODM: original design and manufacturing suppliers.

OFDM: orthogonal frequency division multiplexing.

OFDMA (orthogonal frequency division multiple access): The data processing intensive protocol that has been incorporated in leading 4G technology road maps. It is the core technology implemented in the Mobile WiMAX standard.

omni-directional antenna: An antenna that radiates equal power in a 360° beam pattern. See also **directional antenna**.

PA: power amplifier.

panel antenna: A directional antenna made up of several driven elements mounted in front of a flat reflecting element. It has a flat plastic or fiberglass cover that gives the antenna a panel-like appearance.

PAR: peak-to-average ratio.

parabolic antenna: A directional antenna made up of a driven element plus a parabolic-shaped reflector. The reflector may either be solid metal, metalic rods, or a metalic mesh.

patch antenna: A directional antenna that is a smaller version of a panel antenna. A patch antenna is most often used indoors. It is often embedded into portable user devices. See also **panel antenna**.

path loss: The attenuation of a wireless signal as it travels from transmitter to receiver. It consists of free-space path loss plus additional attenuation from collisions with obstructions located within the Fresnel zone. See also **attenuation**, **Fresnel zone**.

PBCCH (Packet Broadcast Control Channel): In GPRS systems, this is the broadcast-only channel that alerts user devices of incoming traffic.

PCRF (policy control and charging rules function): It is used to determine and enforce dynamic QoS and charging policies to all the network infrastructure elements in a wireless communications system.

PCS: personal communications service.

PHS: personal handyphone system.

PHY: physical layer.

PDA: personal digital assistants.

PDN: packet data network.

PoC: push-to-talk over cellular.

polarization: The orientation relative to the earth of a wireless signal as it leaves the antenna. It may be vertical, horizontal, or circular. The polarization of the signal changes when the signal reflects off an object.

PTT: push-to-talk.

PW-LANs: public wireless LANs.

QAM: quadrature amplitude modulation.

QoS (quality of service): Describes a wide range of technologies that are employed in wireless and packet switched networks to ensure the delivery of information rates and contractual terms of service quality.

QPSK: quadrature phase shift keying.

RAB (radio access bearer): A nonstandardized term used in UMTS and 3G systems to control the resources required to transfer data between the user equipment and the core network.

RADIUS (Remote Authentication Dial-In User Service) server: RADIUS describes a solution that manages the AAA (authentication, authorization, and accounting) requirements of IP-based networks.

RAM: random access memory.

RAN: radio access network.

RAS: radio and antenna subsystem.

RBS: radio base station.

RF: radio frequency.

RLC: radio link controller.

RNC: radio network controller.

ROHC: robust header compression.

ROM: read-only memory.

RSS: Really Simple Syndication.

RTLS: real-time locating systems.

RTP: Real-Time Transport Protocol.

RTSP: Real-Time Streaming Protocol.

RUIF: radio unit interface.

RUS: Rural Utility Service.

RX: receiver.

RXIF: receiver interface.

RXRF1: receiver RF1.

RXRF2: receiver RF2.

SC-FDMA: single-carrier frequency division multiple access.

SCDMA: synchronous CDMA.

SAE: system architecture evolution.

SAN: storage area network.

SARFT: State Agency for Radio, Film, and Television.

SAS: software as (a) service.

SCP: Spectral Compression Positioning™

SDMA: space division multiple access.

SDP: Session Description Protocol.

SDR: software-defined radios.

sector: A wireless system that serves more than one coverage area from the same antenna site. In cellular systems, there are most often three "sectors" of 120° beamwidth on each cell site. Each sector is created by its own directional antenna system.

sensitivity: The ability of a wireless receiver to detect and successfully decode an incoming wireless signal.

selectivity: The ability of a wireless receiver to discriminate between a wireless signal on the desired frequency and other wireless signals on other frequencies.

SGSN: serving GPRS support node.

SIC: successive interference cancellation.

SIM: Subscriber Identification Module. The card inserted into GSM handsets to manage authentication and authorized features.

SIMO: single input multiple output. See also **MIMO**.

SIP: Session Initiation Protocol.

SISO: single input single output. See also **MIMO**.

SLS: selective laser sintering.

smart antennas: A wide range of technologies are covered under the broad term of "smart antennas." Various types of smart antennas exist including the following technologies: phased array, active electronic beam forming, and antenna nulling techniques, and mechanically steered antennas.

SMS: short message service.

SNR: signal-to-noise ratio.

SOC: system on a chip.

StiMi (satellite terrestrial interactive multimedia infrastructure): The Chinese domestic standard for video broadcasting to cellular type handsets and small devices. Taking its name from the company that developed the technology; TiMi Tech Co. Ltd.

SWR–(standing wave ratio): It indicates the relative efficiency of an antenna system. The lower the SWR, the more power an antenna radiates and the better the wireless link performs.

TCA: telecom-computing architecture.

TCH: traffic channel.

TCP: Transport Control Protocol.

TDD (time division duplex): The use of the time domain to manage both upstream and downstream data transmission over a single radio frequency channel. Fast switching transmitters and receivers allow for these systems to only require about 3% overhead.

TDMA: time division multiple access.

TD-SCDMA: time division-synchronous code division multiple access.

TD-CDMA: time division-code division multiple access.

TIA/EIA: Telecommunications Industry Association/Electronics Industry Association.

TISPAN: Telecoms and Internet Converged Services and Protocols for Advanced Networks.

ToIP: television over IP.

TOR: transmit observation receiver.

TRX: transceiver.

TTI: transmission time interval.

TX: transmitter.

UE: user equipment.

UMA: Unlicensed Mobile Access.

UMB: Ultra Mobile Broadband.

UMTS: Universal Mobile Telecommunications System.

UPCS: Unlicensed Personal Communications Services.

UTRAN: UMTS Terrestrial Radio Access Network.

VDSL: very high speed DSL.

VoIP: Voice over Internet Protocol.

VPN: virtual private network.

WAP: Wireless Application Protocol dedicated to separation between the upstream and downstream transmissions.

wavelength: The physical length of one cycle of the wireless signal. Every wireless signal has both a specific wavelength and a specific frequency. The higher the signal frequency, the shorter the wavelength.

WCDMA: wideband CDMA.

WiBro: Wireless Broadband.

WiFi (Wireless Fidelity): The wireless local area network standard.
WiFi—IEEE standard 802.11(b), (g),(a),(i),(n).
WiFi Alliance.

WiMAX (Worldwide Interoperability for Microwave Access): Describes a family of standards defining a wide range of architectures including point-to-point links, point-to-multipoint fixed services, and a fully mobile wireless broadband version. The evolution of the 802.16 standard and its interoperability and brand certification organization, the WiMAX Forum, has been highly controversial, especially from the legacy cellular industry participants. It has now gained global regulatory parity with the GSM and CDMA camps, as the ITU has now included Mobile WiMAX as one of the approved technologies under the IMS 2000 list of approved standards.
Mobile WiMAX—IEEE standard 802.16(e) OFDMA-TDD.
WiMAX 2004—IEEE standard 802.16(a) OFDM-TDD-TDMA.
WiMAX Forum—Interoperability certification.

WLAN: wireless local area network.

WMAN: wireless metropolitan area network.

WPS: wireless protocol stack.

WRC-07: World Radiocommunication Conference 2007.

Yagi: A directional antenna made up of a "driven element" connected to the transmission line, a "reflector" (signal-reflecting element), and one or more "directors" (signal-directing elements).

B

A SCENARIO OF A BROADBAND WIRELESS CUSTOMER, CIRCA 2012

2012 SCENARIO

A 32-Year-Old Corporate Worker in 2012

Kids grow up. The younger generation is a very connected, online savvy, heavy-media crowd that will eventually replace us old geezers at the workplace. Companies that do not adapt to this new paradigm may discover that it is too late to adapt, and that some new company is going to disrupt their status quo by hijacking all their young workers. But what might the mobile phone look like in this new world?

Let's follow the day in the life of Sunil Jain, a director of design for Calty Design Research (Toyota, Lexus, U.S. design center), headquartered in Newport Beach, California. He has a new messaging hub phone with Universal Broadcast Modem capabilities, connected to 5G bandwidth, 500-GB hard drive, and global voice capabilities with VoIP access on WLAN networks. He has gone to the "dark side" with this e-mail-friendly phone, but uses it mostly for mobile IM and messaging.

6:00 a.m.—Orange County, California airport. Sunil has some time before his flight takes off to the Detroit Auto Show for a new model launch. The pressure of the launch day is on. Using a secure, paid-for airport network, he checks and responds to critical messages and IMs his team in Japan. The IMs lead him to his personalized portal, where he notices some new competitor rumors. It looks like as if the company's "first to market" new turbo-diesel hybrid electric launch will directly compete with an unanticipated Mercedes launch with the same type of

Wireless Broadband. By Vern Fotheringham and Chetan Sharma
Copyright © 2008 the Institute of Electrical and Electronics Engineering, Inc.

power plant. He sends a sneak photo to his team about the announcements and schedules a conference call in two hours with his top lieutenants. For the flight, he downloads, from his music subscription service, the best of four new Kings of Leon songs to relax. He sends a phone movie love note to his wife from his phone with her favorite (DRM-free) track as a gift for the day. When he gets the track, he sees some recommended picks he likes and adds them to the rental list. She gets the song and a promotional offer for a concert ticket and a discounted digital album for a new band that is in the genre that she just received. Out of curiosity, he checks the weather in Detroit and gets an interesting ad asking him to go to a special booth in Cobo Hall, on his arrival, to get a pass for another announcement from Lexus. It is nice to know his marketing group is fully behind this launch!

7:00 a.m.—Sunil's flight is delayed and the gate has changed. LBS on his phone guides him through the shortest route to the new gate. At the airport, he notices a 3D billboard ad for a hot new 52″ OLED television and snaps a photo of the barcode to get more information sent to him. He responds to the billboard by texting "info" to the four-digit SMS code on the ad, immediately receiving a download of the relevant information to his phone, e-mail, and personal Web site for his review. The information is personalized to him on the basis of his mobile's CLI. Now aware of his credit history and his previous purchases, the company offers him a "special" price for immediate purchase and delivery. He also sets up a special auto show version of a Twitter-like rumor site to keep abreast of all the rumors and the buzz. He uses his mobile phone PIM and checks the online presence of team members in LA, Japan, and Detroit for a conference call. The call is arranged through mobile IM from the plane. He then takes a short break and listens to new music from his media phone with noise-canceling headsets. The phone sends a warning vibration to avoid the sluggish looking bagels being handed out, disguised as food (just kidding).

7:10 a.m.—Conference call from his mobile phone. A conferencing application on Sunil's media phone connects three team members into a conference call. They discuss their response to the competitive launch to the press, partners, and dealers. The conference call is digitized, converted into text, and filed into archives by the time they are done with the call.

8:00 a.m.—Sunil watches a video stored on his phone, which is streamed to the video screen on the back of the airplane seat in front of him. Ads are served at the bottom of the screen and then fade out as the movie starts. As the movie plays, he keeps receiving schedule and press-briefing updates from his PR firm on his phone from the network connection.

Before boarding the plane, Sunil checks his alerts to make sure his son made it to school safely and tunes into his "nanny cam" to wave hello to the babysitter and the newborn.

4:00 p.m.—Sunil gets in the limo that picks him up at the Detroit airport and speaks into his phone: "Get us to Cobo Hall in the shortest time possible." The mapping application fixes his location with GPS and real-time, predictive traffic forecasts hit the map. The ads seem to know that he is headed to Cobo Hall for the auto show and express special offers for a five-star dinner at a restaurant, which was previously filled for the evening that Sunil wanted to go to. He clicks on the offer to call and makes reservations for a table. It would be easy to get distracted with map mashups rich with neighborhood information while his driver speeds by the pothole-ridden gang turf out the side window. But then he gets a message outlining the map of the floor of the auto show instead, complete with walking directions to the massive show hall and ads for food in the hall. The neighborhood map, where the restaurant is located, is sent to him in 3D and he can fly around it from his phone. He stores the maps on his media device for later use. Next, he gets a bar code pass for another after-show party on his phone and saves that on his phone as well.

4:30 p.m.—The phones in the conference room don't work—not to worry. A mobile conference call is set up from a pressroom at the hall, with the design office in Newport Beach and the headquarters in Japan to prepare for the evening's launch. Just before the call, Sunil remotely checks the PC he had left at the office for the latest presentation file and pulls it onto his phone. For better resolution, he projects the launch video preview to the wall from his phone. He shows the video of the new design that will debut at 6:00 p.m. at the show to an embargoed press group. They are impressed with the phone's display as well as the new car! Sunil's boss is also on his mobile phone, stuck in traffic on the way to the hall, without his predictive traffic maps, but the video is appropriately transcoded for both mobile devices and bandwidth. After the call, the executives sign off on the new messages for the competitive shifts at the launch. They are sent out via e-mail, TXT, and IM.

5:30 p.m.—Sunil takes a deep breath and passes on the good news to the launch team. It is all go for the launch. He makes a few phone calls from his mobile, prepping last-minute launch details to the team's logistics lead.

6:00 p.m.—The launch goes off with a bang and without a glitch! Sunil takes videos and still photos from his media phone as the event unfolds. He has assigned a team lead to do the same at the Mercedes launch, but he gets lost with a booth model … only to return just in time. Both the Lexus and Mercedes launches are well documented with mobile media devices, professional cameras, and videos. The mobile media devices, however, post automatically to mobile-friendly blogs as the hall's WLAN networks are jammed. Sunil gets a few minutes to update his 6:10 p.m. blog entries from his mobile phone with voice podcasts. It is posted in near real time, complete with video and high resolution still photos taken on the show floor as the announcements were unfolding. Ads are served up automatically in context to the new text on the new postings for readers. As he is leaving the hall, he needs to find a public restroom and asks his phone, "Find me a public restroom nearby," and gets a walking map in 3D with the nearest location—it was the best use of the phone all day!

8:00 p.m.—Sunil enjoys reading the press briefings that he has pulled up on his phone for the industry analysts he is about to meet for dinner. After dinner, Sunil uses his thumbprint-secured phone payment capabilities to pay for the bill, which is automatically appended to his travel timesheet and expenses.

11:00 p.m.—Sunil heads back to the Hilton, where his travel points are automatically updated as he enters the hotel. After video conferencing from his mobile with his kids back home in Newport Beach, he relaxes with a glass of wine and responds to some TXT messages, reminding him of his old style college TXTing with his wife, way back in 1999, and calls it a day.

C

SPECTRUM TABLES—WIRELESS BROADBAND

MOBILE NETWORK SPECTRUM ALLOCATIONS

The majority of the spectra allocated for mobile wireless services have been organized in paired blocks of spectrum for separate radio pathways for both transmit and receive requirements. Thus, the migration of the legacy cellular networks and most new UMTS spectrum licenses to next-generation OFDMA-based LTE technology will be accomplished using FDD techniques. The spectrum that is coming into service under the Mobile WiMAX platforms is typically targeted for implementation using unpaired single blocks of spectrum that allow both transmit and receive signals to travel over the same radio frequencies. The use of TDD techniques allows two-way mobile communications over these unpaired spectrum blocks that were traditionally allocated for broadcasting services.

The following table details the majority of the spectrum allocations that are presently under consideration for conversion to next-generation OFDMA-based services. Not included are the original 800 MHz allocations for analog AMPS cellular in the United States. Most have been converted to either TDMA or CDMA digital technology, but it is highly likely that these bands will also eventually be considered for a migration to OFDMA technology by the existing holders of these legacy licenses.

Wireless Broadband. By Vern Fotheringham and Chetan Sharma
Copyright © 2008 the Institute of Electrical and Electronics Engineering, Inc.

Frequency bands for terrestrial IMT-2000

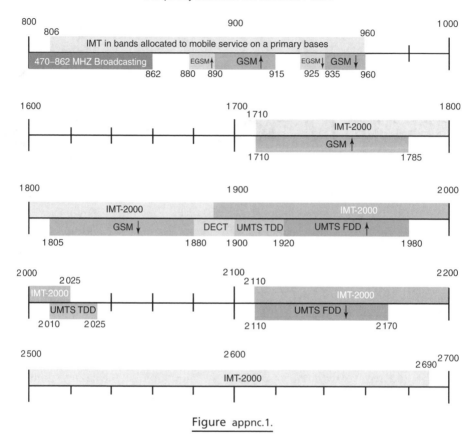

Figure appnc.1.

INDEX

ABOUT THE AUTHORS

Vern Fotheringham is recognized internationally as an industry leader and successful entrepreneur in the wireless and broadband communications industry. Throughout his career, he has been a catalyst for innovation and change in the competitive telecommunications field through direct entrepreneurial activities, as an advisor or influencer on many successful projects, a public policy and regulatory advocate for new telecommunications service rules and standards, and as an inventor and creator of new and innovative services.

Currently, he is the Managing Director of IP Broadband Ltd., in charge of developing converged IP services targeted on the Asian and North American markets for next generation services. He also provides strategic marketing and technical consulting counsel to a number of major telecommunications service providers, including Nextlink Wireless, Inc. where he is the Acting CTO. Mr. Fotheringham is also a managing member of Community Broadband, LLC a consulting firm specializing in advising municipalities on IT and telecommunications matters. Previously he was the President and CEO of ADAPTIX, Inc. a world leader in the development of next generation broadband wireless system technology. ADAPTIX, Inc. pioneered and secured patent protection for the core technology of what is now the IEEE 802.16(e) standard for OFDMA high-speed, scalable, interference immune, metropolitan area IP broadband radio systems.

Highlights of his career activities include: direct participation with the development and international expansion of the cellular telephone industry, working on projects in the Unite States, Canada, Mexico, Germany, Sweden, Hong Kong, Brazil, Argentina, Venezuela and Bangladesh; the creation and development of the mobile satellite industry (Omninet/OmniTRACS, AMSC & NORCOM); earning a U.S. nationwide license for air-to-ground communications services (Claircom/AT&T Wireless); winning the first GSM license in Hong Kong (SmarTone); pioneering in the Digital Audio Radio Service (founder and chairman of Digital Satellite Broadcasting Corporation); and, founding the millimetric microwave industry with the creation of both Advanced Radio Telecom Corporation (ART) as a service provider (Chairman and CEO), and WavTrace a pioneering point-to-multipoint broadband wireless equipment manufacturer (now owned by Harris Corporation). ART was a NASDAQ-listed CLEC and enhanced service provider that held broadband radio spectrum licenses in 207 major U.S. markets, plus five nationwide European licenses. His efforts also resulted in the adoption of the U.S. standards for millimetric microwave regulations and licensing in Japan. Fotheringham pioneered Internet radio and web-based distribution of world music in partnership with Quincy Jones at QRadio. He was also a pioneer in the field of broadband DSL IP access and VoIP hosted services with Bazillion, which was the first nationwide, toll quality VoIP service provider. His early career included positions in public safety with the City of Huntington Beach, California, initially as a Marine Safety Officer, then as a Firefighter and Paramedic.

Vern has also pursued a lifelong interest in specialty automobile manufacturing and motorsports through the creation of Vemac Cars Ltd., a Japanese, U.K. and U.S. international partnership (go to www.vemaccars.com) which was founded to develop green high-performance vehicles. Vemac cars compete in the Super GT Championship racing series in Japan with noteworthy success against the major manufacturers. He was also a pioneer of the single-make racing series concept as the founder of the Formula Mazda (1981 to present) series and the Sports Toyota Championship which was featured as a support race in the early years of the American LeMans series.

Vern is an IEEE member, and received his Bachelor of Arts from California State University, Fullerton. He also pursued graduate studies at the Claremont Graduate School in Claremont, California. He currently serves as a director for a number of early stage ventures.

Chetan Sharma is President of Chetan Sharma Consulting and one of the leading strategists in the mobile industry. Executives from wireless companies around the world seek his accurate predictions, independent insights, and actionable recommendations. He has served as an advisor to senior executive management of several Fortune 100 companies in the wireless space. Sharma has helped several players in the ecosystem develop their mobile advertising strategy. Some of his clients include NTT DoCoMo, Disney, KTF, Comcast, Motorola, FedEx, Sony, Samsung, Alcatel Lucent, KDDI, Virgin Mobile, Sprint Nextel, AT&T Wireless, Reuters, Qualcomm, Reliance Infocomm, SAP, Merrill Lynch, American Express, InfoSpace, BEA Systems, and Hewlett-Packard.

Chetan is the author of four other books on the mobile industry: *Wireless Internet Enterprise Applications* (Wiley, 2000), *VoiceXML* (Wiley, 2002), *Wireless Data Services: Technologies, Business Models, and Global Markets* (Cambridge University Press, 2004), and *Mobile Advertising* (Wiley, 2008). He has patents in wireless communications, is regularly invited to speak at conferences worldwide, and is an active member in industry bodies and committees. Chetan is interviewed frequently by leading international media publications such as *Time* magazine, *New York Times, Wall Street Journal, BusinessWeek, Japan Media Review, Mobile Communications International,* and *GigaOM,* and has appeared on NPR, WBBN, and CNBC as a wireless data technology expert. He addresses several telecommunications industry trade delegations to the United States, such as executive teams from Japan, Korea, and Finland. He served on the U.S. advisory committee of the Korea-Pacific U.S. States Joint Conference on wireless and multimedia. Chetan has published several articles and industry reports on a wide variety of topics.

Chetan is a sought-after strategist on IP matters in the wireless industry. He has advised clients with some of the biggest patent portfolios in the world and has worked with players across the wireless value chain. He has been retained as an expert witness and advisor for some of the most prominent legal matters in front of the International Trade Commission (ITC) including *Qualcomm vs. Broadcom* and *Ericsson vs. Samsung.*

Chetan Sharma is a Senior Member of IEEE, IEEE Communications Society, and IEEE Computers Society. He has Master of Science and Electrical Engineering degree from Kansas State University, Manhattan, Kansas and Bachelor of Science degree from the Indian Institute of Technology, Roorkee, India.